T0177549

The Phantom Pattern Problem

In his earlier life as the very successful author of statistics textbooks, Gary Smith had a knack for creating creative applications that helped students learn important statistical concepts in a fun and intuitive way. In this collaboration with Jay Cordes, Smith takes this same approach mainstream with entertaining example after entertaining example highlighting their central point that not all patterns are meaningful. Smith and Cordes argue that the solution to the dilemma is not more data, but rather more intelligent theorizing about how the world works. Readers should heed their warning—and, those who don't should not be surprised if they make an appearance in the next Smith and Cordes book as a cautionary tale!

Shawn Bushway, Senior Policy Researcher
Behavioral and Policy Sciences Department, RAND Corporation

A nice little antidote to big claims about big benefits of Big Data.

Marc Abrahams, Editor of the *Annals of Improbable Research*
founder of the Ig Nobel Prize ceremony

It's refreshing to see a book on paleo that's about distance running, pattern recognition, and jokes, rather than scarfing steaks, pumping iron, and violence. But, as Gary Smith and Jay Cordes explain and demonstrate, pattern recognition can lead to superficially appealing but ultimately misleading conclusions.

Andrew Gelman, Professor of Statistics and Computer Science
Columbia University

Gary and Jay hit the ball out of the park with "The Phantom Pattern Problem: The Mirage of Big Data." Full of fun stories and spurious correlations and patterns, the book excels at its aim: Explaining the hazards of big data, how many can easily be fooled by putting too much trust in blind statistics, as well as highlighting many pitfalls such as overfitting, data mining with out-of-sample data, over-reliance on backtesting, and "Hypothesizing after the Results are Known," or HARKing. The text is a home run on the importance of building models guided by human expertise, the critical process of theory before data, and is a welcome addition to any reader's library.

Brian Nelson, CFA, President
Investment Research, Valuentum Securities, Inc.

The legendary economist Ronald Coase once famously said, 'If you torture the data long enough, it will confess.' As Smith and Cordes demonstrate in spades, the era of Big Data has only exacerbated Coase's assertion. Packed with great examples and solid research, "The Phantom Pattern Problem" is a cri de coeur to those who believe in the unassailable power of data.

Phil Simon
Award winning author of *Too Big to Ignore: The Business Case for Big Data*

Using easily understood examples from sports, the stock market, economics, medical testing, and gambling, Smith and Cordes illustrate how data analytics and big data can be seductively misleading. I learned a lot.

Robert J. Marks II, Ph.D.
Distinguished Professor of Electrical & Computer Engineering, Baylor University
Director, The Walter Bradley Center for Natural & Artificial Intelligence

THE PHANTOM PATTERN PROBLEM

The Mirage of Big Data

GARY SMITH AND JAY CORDES

OXFORD
UNIVERSITY PRESS

OXFORD

UNIVERSITY PRESS

Great Clarendon Street, Oxford, OX2 6DP,
United Kingdom

Oxford University Press is a department of the University of Oxford.
It furthers the University's objective of excellence in research, scholarship,
and education by publishing worldwide. Oxford is a registered trade mark of
Oxford University Press in the UK and in certain other countries

First Edition published in 2020

Impression: 2

Published in the United States of America by Oxford University Press
198 Madison Avenue, New York, NY 10016, United States of America

British Library Cataloguing in Publication Data

Data available

Library of Congress Control Number: 2020930015

ISBN 978–0–19–886416–5

Printed and bound by
CPI Group (UK) Ltd, Croydon, CR0 4YY

This book is dedicated to our families.

TABLE OF CONTENTS

Surely You Jest

I n October 2001, Apple's Steve Jobs unveiled the iPod, a revolutionary hand-held music player with a built-in hard drive: "To have your whole CD library with you at all times is a quantum leap when it comes to music. You can fit your whole music library in your pocket." Despite the $399 price tag, sales were phenomenal, hitting thirty-nine million in 2006, before being eclipsed by the 2007 introduction of the iPhone.

Figure I.1 shows that the explosion of iPod sales in 2005 and 2006 coincided with an increase in the number of murders in the United States. Were people killing each other in order to get their hands on an iPod? Were iPod listeners driven insane by the incessant music and then murdering friends and strangers?

When we showed Figure I.1 to a friend, her immediate reaction was, "Surely you jest." We are jesting, but there is a reason why we jest. In 2007, the Urban Institute, a highly regarded Washington think-tank, released a research report on the increase in murders in 2006 and 2007:

The rise in violent offending and the explosion in the sales of iPods and other portable media devices is more than coincidental. We propose that, over the past two years, America may have experienced an iCrime wave.

Unlike us, they were not jesting.

We have all been warned over and over that correlation is not causation, but too often, we ignore the warnings. We have inherited from our distant ancestors an often-irresistible desire to seek patterns and succumb to their allure. We laugh at some obviously nutty correlations; for example, the number of lawyers in Nevada is statistically correlated with the number

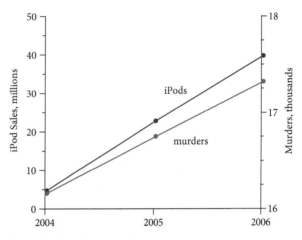

Figure I.1 iPod sales and murders.

of people who died after tripping over their own two feet. Yet other correlations, like iPod sales and murders, have a seductive appeal. If esteemed researchers at the Urban Institute can be seduced by fanciful correlations, so can any of us.

Thousands of people didn't kill each other so that they could steal their iPods, and thousands of iPod listeners weren't driven murderously insane. Murders and iPod sales both happened to increase in 2005 and 2006, as did many other things. The serendipitous correlation between murders and iPod sales did not last long. Murders dropped in 2007 and have fallen since then, even though iPod sales continued to grow for a few more years until they were dwarfed by iPhone sales.

The correlation between murders and iPod sales is particularly laughable since it is based on a mere three years of data. Anything that increased (or decreased) steadily during this two-year span will be highly correlated with murders—for example, ice cream sales in the U.S.

Figure I.2 shows that the correlation between ice cream sales and murders is as good as the correlation between iPod sales and murders. Were people killing each other in order to snatch iPods or ice cream cones? Almost surely neither, and yet, two experienced researchers at the Urban Institute assumed that correlation was causation.

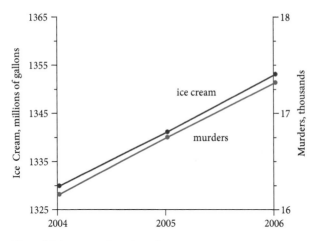

Figure I.2 Ice cream sales and murders.

How, in this modern era of big data and powerful computers, can experts be so foolish? Ironically, big data and powerful computers are part of the problem.

We have all been bred to be fooled—to be attracted to shiny patterns and glittery correlations. Big data and powerful computers feed this addiction because they make it so easy to find such baubles—and they also ensure that most of what we find is as worthless as the fanciful claim that iPods increased the murder rate.

It is up to us to resist the allure, to not be fooled by phantom patterns.

CHAPTER 1

Survival of the Sweaty Pattern Processors

Compared to other animals, humans are not particularly strong or powerful. We don't have sharp teeth, claws, or beaks. We don't have sledgehammer horns, tusks, or tails. We don't have body armor. We are not great swimmers or sprinters. How did our distant ancestors not only survive, but become masters of the planet?

Two things powered our ascent: cooling efficiently and recognizing patterns.

Survival of the Sweatiest

Our first great advantage is that we are born to run. Our prehistoric ancestors walked and ran everywhere, chasing things to eat and running away from beasts that wanted to eat them. Humans are not fast, but we do have exceptional endurance. The early humans who could run down their prey by wearing them out ate plenty of meat, which made them stronger and more likely to mate and pass along their genes to their children. Through countless generations of chasing, feasting, and mating, humans who couldn't keep up were weeded out of the gene pool.

Hunting based on endurance running is called *persistence hunting*— pursuing prey until they cannot run any more. Part of our success is due to how our bodies evolved two million years ago in ways that facilitate endurance running, long before our more recent ancestors developed arrows, spears, and other projectile weapons. Among the adaptations identified by Harvard Anthropology Professor Daniel Lieberman:

We developed long, springy tendons in our legs and feet that function like large elastics, storing energy and releasing it with each running stride, reducing the amount of energy it takes to take another step. There are also several adaptations to help keep our bodies stable as we run, such as the way we counterbalance each step with an arm swing, our large butt muscles that hold our upper bodies upright, and an elastic ligament in our neck to help keep our head steady.

Even the human waist, thinner and more flexible than that of our primate relatives, allows us to twist our upper bodies as we run to counterbalance the slightly-off-center forces exerted as we stride with each leg.

The key to successful endurance running is not simply strong legs and good balance (plenty of animals have that), but keeping the body from overheating, and humans excel at this.

Even when exterior conditions fluctuate greatly, the human body is very good at keeping its core temperature in the narrow safe range 36.5°C to 37.5°C (97.7°F to 99.5°F). There is *hypothermia* if a human body temperature drops to a dangerously low level, say 35°C (95°F) and, at the other extreme, there is *hyperthermia* if the body temperature rises to a dangerously high level, say 38°C (100.4°F).

Heat exhaustion, with dizziness, weakness, lack of coordination, and cardiovascular strain that can lead to heat stroke and death, is most often caused by strenuous physical exertion in a hot environment; for example, running or playing active sports on unusually hot days. Fortunately, humans have an excellent built-in cooling system. We are relatively hairless; we can adjust our breathing rate while running; and we sweat more per square inch than any other species. In contrast, hairy animals that cool themselves by panting overheat quickly—which is why many rest at midday, and why humans can outlast them during a midday chase.

Persistence hunting is still practiced in parts of Africa, Mexico, and Siberia. In the 1980s and 1990s, an anthropologist witnessed native hunters in Botswana run down large antelopes. A hunt might last five hours and cover more than twenty miles in temperatures above 100 degrees Fahrenheit. The antelope fled the human hunters, rested when it began to overheat, and then ran again when the humans caught up to it. This cycle repeated until hyperthermia set in and the antelope collapsed because it was too overheated to continue running.

In 2013, the BBC reported that a Kenyan herdsman had run down two cheetahs that had been killing his goats. He didn't shoot them with a gun;

he simply ran them into the ground. He explained that, one morning, "I was sipping a cup of tea when I saw them killing another goat." He waited until midday, and then set off with some youths from his village. After a four-mile pursuit, they tied up the exhausted cheetahs with ropes and turned them over to the Kenya Wildlife Service. Our ancestors would be proud, though puzzled why they didn't eat the cheetahs.

Survival of the Smartest

It turns out that endurance capacity, brain size, and cognitive abilities are all related. The benefits of exercise are not directly hereditary, but individuals whose brain cells and cognitive thinking responded the most to exercise would have an evolutionary advantage over others.

There was generation after generation of natural selection—the survival of the smartest. The smartest thrived and mated and passed along their brain power to their children, among whom the smartest were more likely to survive and mate and have children. The eventual consequences of this virtuous cycle are that, relative to our body size, human brains are now three times larger than the brains of other mammals, and human intelligence far surpasses other animals.

Exercise seems to make us smarter. Personally, we've always believed that just as we exercise our bodies, we should exercise our brains by reading, playing card games, and doing puzzles. The latest research suggests that our brain power may actually be boosted more by exercising our bodies than by exercising our brains!

One interesting experiment put mice in four different living environments for several months. Some mice lived in a simple cage with no toys and ate plain food. The second group lived in colorful cages with mirrors and seesaws and ate gourmet food (by mice standards). The third group had no toys or fancy food, but had a running wheel to exercise in. The fourth group lived the richest lives, with colorful toys and fancy food and a running wheel.

All the mice were given mental challenges before the experiment and afterward. The only thing that made them smarter was the running wheel. Fancy toys and food made no difference, either for the mice who had running wheels or for those who didn't. Exercise consistently made the mice smarter, regardless of whether they had interesting toys and gourmet food or no toys and bland food.

How does it work? For one thing, exercise pumps blood and oxygen to the brain, which nourishes brain cells. Exercise also increases the production of a protein called BDNF, which helps brain cells grow and connect with each other.

In addition to mice running in wheels, numerous experiments with humans have come to the same conclusion. One study gave a group of young men a memory test. Then some of the men sat still for thirty minutes, while others rode stationary bikes at an exhausting pace. When they were retested, the men who sat still performed the same as before. However, the men who exercised had higher BDNF levels and did better on the memory test. Many similar studies have found that exercise helps brain cells grow and survive and improves memory and learning.

It appears to work for all age levels.

A study of teenagers found that they did better on problem-solving tests after exercising thirty minutes a day. A study of middle-aged people found that they had higher scores on tests of memory and reasoning after four months of strength and aerobic training.

A study of people over the age of fifty with memory problems found that they scored better on mental tests after exercising regularly for six months. A study of the elderly found that exercising three times a week reduced the chances of Alzheimer's disease by a third. A study of patients with Parkinson's disease show marked improvement after a stationary bike regimen.

Doctors have long believed that exercise slows down the effects of aging on memory ("Where did I put my keys?") and learning ("How does this new phone work?"). Now it seems that exercise can actually reverse the process, not only slowing the death of brain cells, but also growing new ones. For example, one study found that people who exercised for three months grew new neurons, particularly in a region of the brain responsible for memory and learning.

Pattern Recognition

In comparison to other animals, evolution seems to have favored our distant ancestors who had exceptional endurance and extraordinary brains. Exceptional endurance is good for hunting. What are extraordinary brains good for?

For one thing, our super-sized brains complemented our remarkable endurance. Persistence hunting is enhanced by intelligence, communication skills, and teamwork, which may have fueled the evolution of human intelligence, communication, and social skills. The best human hunters tended to be smarter than average, which enabled them to track down prey, plan strategies for group hunting against bigger and faster animals, and execute these plans. Those who were smarter than average came to dominate the gene pool as those who were less intelligent tended to be less successful hunters, more likely to die young, and less able to attract fertile mates.

It is not just the size of our brains. Many whales, elephants, and dolphins have bigger brains than humans, but how many can write beautiful poetry and inspiring symphonies? How many can design automobiles, and build machines that build automobiles?

A large part what is meant by intelligence is recognizing patterns, and pattern recognition had terrific evolutionary payoffs for humans. Here is a very incomplete list of obviously helpful patterns that had survival and reproductive value for those who recognized them:

- Zebra stampedes signal predators.
- Elephants can be followed to water.
- There is a recurring cycle of night and day.
- There are growing seasons.
- Dark clouds signal rain.
- Tides come in and go out.
- Some foods are edible; others are poisonous.
- Some animals are prey; others are predators.
- Fertile mates can be identified.

The survival and reproductive payoffs from pattern recognition gave humans an evolutionary advantage over other animals. Those who were better able to recognize signs of danger and fertility were more likely to pass on their pattern-recognition abilities to future generations. Indeed, it has been argued that the cognitive superiority of humans over all other animals is due mostly to our evolutionary development of superior pattern processing.

Many of the prehistoric patterns that our ancestors noticed were surely useless. For example, during a solar eclipse, people might chant or dance, and when the sun reappears conclude that the sun god was pleased by the chanting and dancing. Such misinterpretations are understandable, but of little consequence. The strong evolutionary advantage of recognizing useful patterns can easily override the relatively minor costs of being misled by useless patterns.

Humans share some valuable pattern-recognition skills with other animals:

- recognition—identifying one's species and threatening actions.
- signaling—communication via gestures.
- mapping—using landmarks to remember the location of food, shelter, and predators.

The pattern-recognition skills in which humans far surpass other animals include:

- communication—written and spoken languages that convey detailed information.
- invention—the creation of sophisticated tools and other means of achieving specific goals.
- arts—the creation of aesthetically pleasing writing, drawing, music, and sculptures.
- imagining the future—specifying complicated possible consequences of actions.
- magic—a willing suspension of disbelief; entertaining impossible thoughts.

In many ways, intelligence often involves an ability to make good decisions based on the detection of patterns. Some believe that there are different kinds of intelligence, such as being math-smart, word-smart, and people-smart, but all of these skills are enhanced by pattern recognition; for example, recognizing mathematical principles that can be widely applied, recognizing effective ways to string words together, or recognizing the moods and emotions of others. The one thing that these supposedly different types of intelligence have in common is pattern processing.

Some make a distinction between intelligence and creativity but, again, pattern processing is important. *Intelligence* might be defined as using stored information and logical reasoning to determine the correct answer to a question or the most promising course of action. Pattern recognition is clearly required for determining the best answer to a question or identifying the possible consequences of specific actions. *Creativity*, in contrast, might be defined as the consideration of surprising, outside-the-box answers, actions, and consequences. Solutions are intelligent. Jokes are creative. However, creativity benefits from pattern recognition, too.

Knowledge of a pattern can help generate original ideas that don't fit the pattern. Jokes are often funny precisely because the punch line is not what we have become accustomed to expect.

Here is the winner of a funniest-joke-in-the-world contest:

A couple of New Jersey hunters are out in the woods when one of them falls to the ground. He doesn't seem to be breathing and his eyes have rolled back in his head. The other guy whips out his mobile phone and calls the emergency services. He gasps to the operator: "My friend is dead! What can I do?" The operator, in a soothing voice, says: "Just take it easy. I can help. First, let's make sure he's dead." There is a silence, then a shot is heard. The guy's voice comes back on the line. He says: "OK, now what?"

Childish Patterns

Our pattern-recognition abilities manifest themselves at an early age when babies and toddlers use sights and sounds to identify objects and develop language skills. Infants quickly recognize the ways in which dogs, chairs, and water are different, and they learn to anticipate the consequences of different actions and events.

How do toddlers learn to walk? By taking thousands of steps and falling down thousands of times until they figure out the pattern—the combination of actions that keeps them upright as they move about on two legs.

How do children learn to spell and construct sentences? By recognizing patterns. They remember that the sounds *dog* and *cat* are spelled D O G and C A T. Add an *s* when there is more than one dog or cat. Put A N D in between the two words if you want to talk about both dogs and cats. Language acquisition, including spelling, grammar, and all the rest, can be described as pattern recognition, and the fact that humans are so good at pattern recognition is one reason that our language skills are so much more advanced than those of other animals.

Math is more of the same. What is counting, but remembering patterns: 1, 2, 3, and 11, 12, 13? What is arithmetic, but remembering rules? What is computer programming, but applying general principles (patterns if you will) to new tasks?

A friend recently told us that her skinny four-year-old was always too busy to eat. Placed in a chair at the dining room table, he clenched his teeth and struggled to get free—to run around the house to play with anything he could get his hands on. She eventually figured out a way to feed him. When she put him in front of an interesting video, he would open his mouth and receive food. It was like Pavlov's dog experiments: the video went on and the mouth popped open. She recognized, however, that she was the one who had been trained by the pattern—to turn on the video so that he would eat.

One of Gary's children learned at a very young age that saying, "Dis," and pointing at something with both hands caused Gary and his wife to bring over everything in the vicinity of where the child's fingers were aimed. The infant hadn't developed language skills yet. She just knew that the sound, "Dis," accompanied by finger pointing, got results.

Perhaps this child had seen an adult point a finger and make sounds that included "this," followed by something being passed to the person. If it happened more than once, the sequence of events reinforced the pattern in the baby's mind. It was surely reinforced when the baby pointed and said, "Dis," and her parents scrambled to give her what she wanted. The infant recognized the pattern—point and you shall receive—but the parents had also been trained to follow the pattern—fetch and she shall be quiet.

It was certainly a relief for Gary and his wife when the baby figured out more sophisticated patterns by breaking the speech code—learning the specific sounds (i.e. words) associated with the things she wanted and the ability to string words together to form sentences.

Sometimes, a baby's rapid development of pattern-recognition skills can be misinterpreted by adults. For example, it is surely true that a baby who cries and receives attention learns that crying brings attention, and it may well be that babies who figure this pattern out quickly are smarter than average. Some child-development researchers once tabulated the amount of crying by thirty-eight babies when they were four to seven days old, and then measured each baby's IQ at three years of age. The correlation, shown in Figure 1.1, was 0.45, which is highly statistically significant. If crying and IQ were independent, there would be less than a 1-in-200 chance of such a high correlation.

No, this doesn't mean that you can raise a baby's IQ by making it cry! A more plausible explanation is that lively, inquisitive babies want more attention and learn the pattern quickly: crying brings attention.

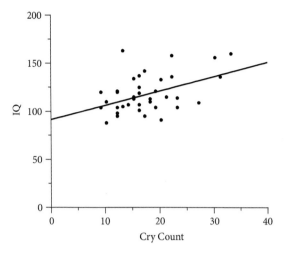

Figure 1.1 Does crying make you smarter?

The Apple Didn't Fall Far From the Tree

Human intelligence relies on pattern recognition—recalling and applying pattern rules stored away in our magnificent brains. Extraordinary intelligence often involves identifying patterns that others don't notice or recognizing how an unfamiliar situation relates to a familiar pattern.

One of the greatest "aha!" moments in the history of science was Sir Isaac Newton's observation of an apple falling from a tree in his mother's garden. Newton had been born on a farm near Grantham, England, and was attending Cambridge University when an outbreak of the bubonic plague closed the university in 1665 and forced Newton to move back to the family farm. (Will the COVID-19 quarantine inspire genius, too?)

Newton later said that he was sitting in an apple orchard when he saw a ripe apple drop to the ground (no, it did not hit him on the head.) Why, he wondered, if the earth is round with no real top or bottom, do apples always fall toward the earth, no matter where the apple tree is on the planet? Not only that, but apples always fall perpendicular to the ground. His explanation of this pattern was that a gravitational force was pulling the apple toward the earth.

At Cambridge, Newton had been trying to understand why the moon orbited around the earth instead of shooting off into space. Now he had an answer. The apple pattern and moon pattern were related! Gravitational forces apply to planets as well as apples and extend over vast distances.

Using data on the observed motions of celestial bodies in the solar system, Newton was eventually able to formulate a remarkably simple law of universal gravitation: every celestial body is attracted to every other body with a force that is proportional to the product of their masses and inversely proportional to the square of the distance between them.

Newton's law of gravity predicts the trajectories of soccer balls, arrows, rockets, and other objects hurled in the air. It also explains the movements of tides and the paths of astronomical bodies.

When Uranus was discovered in 1781, its orbit differed somewhat from that predicted by Newton's law of gravity, unless its orbit was being affected by another, as-yet-undiscovered planet. Sure enough, Neptune was eventually discovered, right where Newton's law predicted it would be.

The pattern Newton observed led him to formulate a powerful rule that helps us understand the world and make useful predictions. Of course, Newton's law of gravity was eventually superseded by Einstein's general theory of relativity, but that's another story (and another pattern).

Patterns Everywhere, Until They Aren't

A pattern is reinforced when it repeats. The more often a toddler sees people sitting in chairs, the stronger is the belief that chairs are made to be sat on. The more often a chair is called a *chair*, the stronger is the belief that the word and object go together.

A pattern can be an arrangement of objects (like two chopsticks or a knife, fork, and spoon for eating) or a sequence of events (like a ball thrown upward that falls back to earth). A pattern can lead us to expect a pattern to continue (like servers in a restaurant passing out menus after seating diners) and to notice abnormalities or outliers that don't fit the pattern. We notice if there is a knife and spoon, but no fork, or if a server does not pass out menus.

Poker players notice their opponents' tells. Sports teams notice tendencies. Tells and tendencies can also be used for deception. A savvy poker player can lure an opponent into thinking that she closes her eyes briefly before she bluffs, and then drop the hammer by closing her eyes with a monster hand.

A football team can go all season without faking a field goal and then fake one in the season-ending championship game. Football players and coaches try to identify player and team tendencies by studying hours and hours of film of previous games. Then, during a game, football defenders may recognize a play from the way the players line up and how certain key players move before and after a play starts. Quarterbacks read defenses the same way. This is why teams watch films of their own games—to identify tendencies their opponents may have identified, and then surprise opponents by breaking these tendencies in crucial situations.

One of the most famous plays in Super Bowl history occurred in Super Bowl XLIX in 2015. The New England Patriots led the Seattle Seahawks 28–24, but Seattle had the ball on the New England one-yard line with twenty seconds left in the game. Seattle's great running back, Marshawn "Beast Mode" Lynch had carried the ball twenty-four times for 102 yards (4.25 yards per carry) and the commentators and fans assumed that Lynch would be given the ball and sledgehammer his way into the end zone to win the game.

However, the Seahawk coaches decided to break the pattern and throw a quick pass to another player after he made a sharp cut into the end zone. If this surprise play had worked, the coaches would have been lauded for their outside-the-box thinking. However, a backup Patriots defender, Malcom Butler, an undrafted rookie who had come into the game in the second half, saw how Seattle lined up (a "two receiver stack formation") and remembered the play that Seattle often ran from this formation. He had tried to defend this play several times in practice (never successfully), but this time he intercepted the pass, giving the Patriots their fourth Super Bowl victory. Butler later said that, "From preparation, I remembered the formation they were in…I just beat him to the route and made the play." It was the first pass that Butler had ever intercepted, and it happened because Seattle tried to break one pattern and Butler recognized the different pattern.

Seattle did not disguise its Super-Bowl-ending play successfully, but this cat-and-mouse game can go either way. In the 2019 NFL Super Bowl, New England quarterback Tom Brady, playing in his nineteenth season (and arguably the greatest NFL quarterback of all time), was repeatedly fooled when the opposing team, the Los Angeles Rams, lined up as if their players were in man-to-man coverage and then switched to a zone defense after the ball was snapped to Brady. It can be useful to base decisions on patterns. It can also be useful to base decisions on how competitors interpret and react to patterns.

Recognizing Faces

Humans are incredible at recognizing faces because we instinctively identify patterns that differentiate one face from another. We expect a mouth, nose, two eyes, and other facial characteristics to be in certain locations and to have certain sizes and shapes. We are quick to identify departures from the general pattern—large ears, dimpled chin, bushy eyebrows—in much the same way that caricaturists emphasize unusual facial features.

These differences between a specific face and the general pattern are called distinguishing features, because it is differences, not similarities, that allow our brains to recognize people in an instant, even if the face is partly obscured by eye-glasses or shadows.

Humans can be confused by faces that depart too far from what we expect. We have trouble recognizing faces if the eyebrows are removed completely, and we are terrible at identifying upside-down faces—because the facial patterns we remember almost invariably involve right-side-up people with eyebrows. We are confused because we know and understand the features that make up faces. We know what eyebrows are and we expect to see them. We know what eyes, noses, and mouths are and we expect to see them in that order, from top to bottom, on a face.

Computer facial-recognition algorithms, in contrast, are very brittle because they use mathematical representations of the pixels that make up computer images, and have no real understanding of what a collection of pixels represents. Computer algorithms do not know what eyes, noses, and mouths are; they just notice changes in pixels as the algorithm scans across a digital image.

This lack of understanding can cause hilarious errors when an algorithm tries to match the pixel patterns in different images. One state-of-the art algorithm misidentified a male computer scientist wearing unusual glasses as a female movie star. A picture of a cat with a few altered pixels was misidentified as a bowl of guacamole. A turtle was misidentified as a rifle.

Deceptive Patterns

Sometimes, our instinctive pattern-seeking desires can seduce us into seeing patterns that are imaginary or meaningless. In 2004 an online casino paid $28,000 for a ten-year-old grilled cheese sandwich (with a bite out of it) that was said to contain the Virgin Mary's image. Images of Jesus

Figure 1.2 Jay's family discovered a presidential biscuit.

have been reported in potato chips and dental X-rays. Mother Teresa was spotted on a cinnamon bun (the *nun bun*).

With enough sandwiches, potato chips, dental X-rays, and cinnamon buns, there are bound to be some unusual shapes that one could imagine look like something real (Figure 1.2).

The Luckiest Baby Alive

Though most 7-Eleven convenience stores are now open twenty-four hours a day, the stores got their name in 1946 when they began operating from 7 a.m. to 11 p.m., selling drinks, snacks, diapers, and other essentials. Their signature drink is a mushy, icy combination of water, carbon dioxide, and flavored syrup that is called a *slurpee* because of the sound it makes when drunk through a straw.

Since 2002, 7-Elevens have been giving away free slurpees on July 11 because the month and day (in American date format) are 7/11. In July 2019, a strange thing happened; indeed, National Public Radio's *Morning Edition* headlined the story as "Strange News." CNN reported the strange news this way:

7-Eleven Day typically means free Slurpees for everyone, but this year's celebration turned out more special than usual for one Missouri family. Rachel Langford, of St. Louis, gave birth to a baby girl July 11 — yes, 7/11. That's not all. Baby J'Aime Brown was born at 7:11 p.m., weighing 7 pounds and 11 ounces.

An NBC affiliate said that J'Aime was the "luckiest baby alive," and "might as well be christened Lady Luck."

Among the avalanche of comments on the story, were three main threads.

- It was extremely unlikely:
  - Wow, talk about a crazy coincidence!
  - Does anyone know the spiritual meaning of this?
- She should get something free from 7-Eleven:
  - A lifetime supplies of 7/11 foods and drinks.
  - She should get free slurpees for life!!
- Use the numbers 7 and 11 when buying lottery tickets:
  - They better play that number every day.
  - Awesome…Let me run to 7-Eleven & play 711.

Our personal favorites, though, are

- Please don't name her slurpee!
- And why is this news?

Why *is* this news? On average, eight babies are born every minute in the United States. There is nothing remarkable about a baby being born on July 11 or being born during any specific minute. Pick a minute, any minute, and it is likely that eight babies were born during that minute. There are two 7:11 times every day, 7:11 a.m. and 7:11 p.m., so we expect sixteen 7:11 babies every day, including July 11.

The reported birth weight of 7 pounds, 11 ounces, makes J'Aime a bit more unusual, but this is not an unusual birth weight. A skeptic might also wonder if the reported birth weight had been rounded up or down a bit in order to get nationwide publicity, and maybe some freebees from 7-Eleven—a skepticism encouraged by news reports that the baby's parents planned to contact 7-Eleven.

Sure enough, 7-Eleven gave the family a gift basket and donated $7,111 to her college fund. Now, maybe the company can track down all the other 7/11 babies and do the same.

How to Avoid Being Misled by Phantom Patterns

You can take the human out of the Stone Age, but you can't take the Stone Age out of the human. Pattern recognition prowess served our ancestors well and is surely a large part of the reason that we have evolved from wimps to masters. Today, we are hard-wired to notice patterns, and this innate search for patterns often helps us understand our world and make better decisions.

However, patterns are not always infallible. Sometimes, they are an illusion (like images of the Virgin Mary on a grilled cheese sandwich). Sometimes, they are a meaningless coincidence whose importance we exaggerate (like giving birth at 7:11 on July 11). Sometimes, they are harmful (like buying lottery tickets because we think that a serendipitous pattern predicts the winning number).

Our ancestors discovered patterns in the physical world by using their five basic senses: touch, sight, hearing, smell, and taste. We are now tossed and turned by a deluge of data, and those data are far more abstract, complicated, and difficult to interpret than the information processed by our distant ancestors. Today, most information we receive and process is a digital representation of real phenomena, like data on income, spending, crime rates, and stock prices. And increasingly, our pattern searches are turned over to computer algorithms that know nothing about the real world. The number of possible patterns that can be identified relative to the number that are genuinely useful has grown exponentially—which means that the chances that a discovered pattern is useful is rapidly approaching zero. We can easily be fooled by phantom patterns.

CHAPTER 2

Predicting What is Predictable

The French anthropologist, Claude Levi-Strauss, witnessed a massacre in the Brazilian jungle in the 1930s that happened because a pattern was misinterpreted:

[A Nambikwara Indian] with a high temperature presented himself at the [Protestant] mission and was publicly given two aspirin tablets, which he swallowed; afterwards he bathed in the river, developed congestion of the lungs and died. As the Nambikwara are excellent poisoners, they concluded that their fellow-tribesman had been murdered; they launched a retaliatory attack, during which six members of the mission were massacred, including a two-year-old child.

This was a disastrous confusion of correlation with causation. No matter how many times we are told that correlation is not causation, our inherited love of patterns seduces us (all too often) into thinking that a meaningless pattern is meaningful.

What is Causation?

What do we mean by causation? A cause-and-effect relationship means that one thing (the *cause*) influences the other (the *effect*). Kicking a soccer ball causes the ball to move. The strength of the cause often influences the strength of the effect. Kicking the ball with more force causes it to move farther.

If we want to predict how far the ball goes, we should consider other factors as well, including the weight of the ball and the wind conditions.

There may be multiple causal factors and effects. The weight of the ball and the wind do not cause the ball to move, but they do affect the distance and direction the ball moves after it is kicked.

Nor do useful relationships have to be 100 percent perfect. When scientists conclude that "smoking tobacco causes lung cancer," they do not mean that literally everyone who smokes a cigarette will develop lung cancer a short time later. Instead, they mean something along the lines of "a man who smokes, on average, more than five cigarettes a day is 100 times more likely to develop lung cancer during his lifetime than is a male non-smoker."

It might be useful to use a word other than *causal*, or to use a broad definition of *causal* to encompass these various reasonable interpretations. No matter how we label it, the crucial distinction is between statistical correlations that are coincidental and statistical correlations that are meaningful because there is an underlying reason for the correlation.

If A is correlated with B, there are several possible explanations:

1 A causes B. Rich people tend to spend more because they have more money to spend, not because spending more money makes people rich.
2 B causes A. Stock prices tend to increase when a company announces an unexpectedly large increase in earnings because higher earnings make a company's stock more valuable, not because higher stock prices cause a company's profits to go up.
3 A causes B and B causes A. Cricket players have good hand-eye coordination because good hand-eye coordination makes them better players, and playing cricket improves hand-eye coordination.
4 Something else causes A and B. Students who get high scores on one math test tend to get high scores on a second math test, not because one score affects the other, but because both scores reflect their math ability.
5 It is a coincidence. Murders and iPod sales both increased in 2005 and 2006.

As a practical matter, when we notice a pattern, the relevant question is usually whether there is an underlying reason for the pattern that can be expected to persist so that it can be used to make reliable predictions. A pithy adage is:

In order to predict something, it has to be predictable.

For the first four types of patterns enumerated above, the answer is *yes*, there is a real causal structure that allows us to make useful predictions.

For the fifth type of pattern, the answer is *no*. A does not cause B; B does not cause A; and there is no C that causes A and B.

It might be better to use the word *meaningful* instead of *causal* to clarify that when we say that there is a causal relationship, we are not restricting ourselves to A causes B. A meaningful pattern has an underlying causal explanation and can be used to make useful predictions. A meaningless pattern is a phantom pattern that is coincidental and has no predictive value.

For example, the ancient Egyptians noticed that the annual flooding of the Nile was regularly preceded by seeing Sirius—the brightest star visible from earth—appear to rise in the eastern horizon just before the sun rose. Sirius did not cause the flooding, but it was a useful predictor because there was an underlying reason: Sirius rose before dawn every year on July 19 and heavy rains beginning in May in the Ethiopian Highlands caused the flooding of the Nile to begin in late July.

Yellow Pebbles

Imagine that thousands of small pebbles of various sizes, shapes, colors, and densities are created by a 3D printer that has been programmed so that the characteristics are independently determined. A researcher studying a sample of 100 of these pebbles might discover that, coincidentally, the yellow pebbles in this sample happen, by luck alone, to be bumpy more often than are pebbles of other colors.

When the researcher collects a second sample of pebbles created by this 3D printer, it is unlikely that yellow will be a good predictor of bumpiness since the 3D printer determines color and shape independently. The correlation between yellow color and bumpiness in the initial sample was just a coincidence and, so, it vanished.

Now suppose, instead, that some pebbles are found at the bottom of a lake, and that there *is* some scientific reason why bumpy pebbles in this lake tend to be yellow. In this case, the correlation between bumpiness and yellowness is a useful predictor, even if we don't completely understand why bumpy pebbles are often yellow. It may be that yellow pebbles tend to be bumpy because something living in the lake is attracted to yellow pebbles and likes to nibble on them. Or it may be that bumpy pebbles are more hospitable to the growth of yellow algae. Or maybe a certain kind of soft rock happens to be yellow and bumpy. The crucial distinction between this scenario and the random 3D printer is that the pattern is meaningful because there is an underlying causal reason why bumpy pebbles are often yellow. We don't need to know precisely what the cause is, but it is the existence of a real reason that makes this a meaningful relationship that is useful for making predictions. Otherwise, it is a fragile, useless pattern.

Prediction

Some argue that if prediction is the goal, we don't need causation. Correlation is enough. If there is a correlation between Facebook *great!*s and heart attacks, there doesn't need to be a causal explanation. It is enough to know that they are correlated because one predicts the other. The problem with this argument is that if there is no logical reason for a pattern, it is likely to be a temporary, spurious correlation—like random pebbles created by a 3D printer and like heart attacks and Facebook *great!*s.

One correlation enthusiast gives this example:

To see the difference between prediction and causal inference, imagine that you have a data set that contains data about prices and occupancy rates of hotels…Imagine first that a hotel chain wishes to form an estimate of the occupancy rates of competitors, based on publicly available prices. This is a prediction problem…[H]igher posted prices are predictive of higher occupancy rates, since hotels tend to raise their prices as they fill up (using yield management software). In contrast, imagine that a hotel chain wishes to estimate how occupancy would change if the hotel raised prices across the board…This is a question of causal inference. Clearly, even though prices and occupancy are positively correlated in a typical dataset, we would not conclude that raising prices would increase occupancy.

For predictions to be useful, they must be reliable with fresh data, and consistently reliable predictions with fresh data require a causal structure. In this hotel example, the statistical correlation between prices and occupancy rates is not a fluke; it reflects a real underlying relationship.

It is, of course, possible to misinterpret a relationship between A and B as A causes B when, in fact, it is B that causes A. When it rains, people walk around with open umbrellas above their heads. This is because rain causes people to put up umbrellas, not because putting up umbrellas makes it rain. In the same way, increased demand for hotel rooms tends to

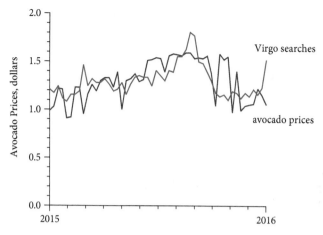

Figure 2.1 Predicting avocado prices from Virgo searches.

cause prices to go up, but raising prices does not increase demand. Umbrellas are a useful sign of rain and increased hotel prices are a useful sign of filled rooms, but we need to be careful in interpreting the direction of causation.

In contrast, Figure 2.1 shows that there is a statistical correlation between avocado prices in San Francisco in 2015 and Google searches for the Virgo zodiac sign, a statistical relationship so strong that there is only a one in 10,000 chance that they would be so closely related by chance alone. There is no underlying reason for avocado prices to be related to Virgo searches, so it was a fleeting coincidence that is useless for making predictions—either of Virgo searches or avocado prices.

Gary recently received an e-mail from a person working for a Chinese information technology company (edited slightly for clarity):

My job is to deal with data and predict the future. Usually, I use a good algorithm such as machine learning. All my predictions are from historical data, but it seems that history data are not very reliable for predicting coming events for companies. Can you think of a better algorithm that will lead to better predictions?

The problem with this request is the belief that the reason why unearthed patterns vanish right when you need them is that there is some problem with the computer program. No. The problem is that there needs to be a real reason underlying the pattern, and no computer program can tell us that. Computers can use numbers to calculate other numbers, but they do not understand what the numbers mean. In the Avocado/Virgo example, computer algorithms do not know, in any meaningful sense, what avocados are and what Virgo searches are, so they have no way of assessing the plausibility of a relationship between avocado prices and Virgo searches. A different algorithm would not make better predictions of avocado prices based on Virgo searches. To make useful predictions, we need to look at meaningful relationships.

Measure Twice, Cut Once

A data scientist employed by a private college was asked to figure out why a substantial number of students left the college before they completed their degree requirements. She poked around in the data and discovered that the financial aid package was a really good predictor—students who applied for aid but didn't get any were very likely to leave the college within the first two years.

Financial aid is discounted pricing. If a college normally charges $50,000 and gives a student $20,000 in financial aid, this is a forty percent discount. The college still gets $30,000 (more if the financial aid is a loan).

This data scientist made some ballpark calculations of the aid packages that could have been offered these students in order to keep them from dropping out while maximizing the college's revenue. When she showed her analysis to the dean of admissions, the dean happened to mention that when a student leaves the college, the dean's office goes into the college's database and resets the student's aid package at zero since the college doesn't give aid to students who have left the college.

Oops! There was a causal explanation, but the interpretation was backwards. It wasn't that zero financial aid caused students to leave. It was that leaving the college caused financial aid to go to zero. Sometimes, it is important to get the correct causal explanation. Assuming that correlation is enough can be an expensive mistake.

Limeys Cured of Scurvy by Limes

Scurvy is a ghastly disease that is now known to be caused by a prolonged absence of vitamin C from the diet. It was once a leading cause of death for sailors who went on long voyages without eating fruits or vegetables containing vitamin C.

It was mostly because of scurvy that Vasco da Gama lost sixty-eight percent of his crew during his 1497–1499 voyage from Portugal to India; Magellan lost ninety percent during his 1519–1522 journey from Spain to the Philippines; and Englishman George Anson lost sixty-five percent during the first ten months of his 1740–1744 circumnavigation of the world. Overall, it has been estimated that more than two million sailors died from scurvy between 1500 and 1800.

For centuries, the conventional wisdom was that scurvy was a digestive disorder caused by many things, including sailors working too hard, eating salt-cured meat, and drinking foul water. Among the recommended cures were the consumption of fresh fruit and vegetables, white wine, sulfate, vinegar, sea water, beer, and various spices.

In 1601 an English trader and privateer named James Lancaster led a fleet of four ships on a five-month voyage from England to southern Africa. The sailors on his ship received daily sips of bottled lemon juice and arrived in good health, while the sailors on the other three ships were given no lemon juice and most contracted scurvy. Lancaster was convinced of the power of lemon juice and dispensed it on his future voyages, but others were skeptical and did not.

In 1747, an English doctor named James Lind did an experiment while on board the HMS Salisbury. He selected twelve scurvy patients who were "as similar as I could have them." They were all fed a common diet of water-gruel, mutton-broth, boiled biscuit, and other unappetizing sailor fare. In addition, Lind divided the twelve patients into six groups of two so that he could compare the effectiveness of six recommended cures. Two patients drank a quart of hard cider every day; two took twenty-five drops of sulphate; two were given two spoonfuls of vinegar three times a day, two drank seawater; two were given two oranges and a lemon every other day; and two were given a concoction that included garlic, myrrh, mustard, and radish root.

Lind concluded that:

The most sudden and visible good effects were perceived from the use of oranges and lemons; one of those who had taken them, being at the end of six days fit for duty…The other was the best recovered of any in his condition; and…was appointed to attend the rest of the sick.

Unfortunately, his experiment was not widely reported and Lind did little to promote it. He later wrote that, "The province has been mine to deliver precepts; the power is in others to execute."

The medical establishment continued to believe that scurvy was caused by the digestive system being disrupted by the sailors' hard work and bad diet, and could be cured by "fizzy drinks" containing sulphuric acid, alcohol, and spices.

In an odd coincidence, in 1795, the year after Lind died, Gilbert Blane, an English naval doctor, persuaded the British navy to add lemon juice to the sailors' daily ration of rum and, thereafter, scurvy virtually disappeared on British ships. (It took another 100 years for the widespread consumption of vitamin C to eradicate scurvy on land.)

Lind's experiment is noteworthy because it was an early example of evidence-based analysis—a serious attempt to conduct a rigorous clinical trial in order to evaluate the efficacy of recommended mediations. It is also instructive to identify the ways in which his experiment could have been improved. For example:

1 There should have been a control group that received no special treatment. Lind's study found that the patients given citrus fared better than those given seawater, but maybe that was because of the ill effects of seawater, rather than the beneficial effects of citrus.

A group of college students once reported seventy-three percent fewer colds than the year before after they had been given an experimental vaccine. This would have been astonishing, except for the fact that a control group reported a sixty-three percent decline!

2 The distribution of the purported medications should have been determined by a random draw. Maybe Lind believed citrus was the most promising cure and subconsciously gave citrus to the healthiest patients.

A study once found that children who were given an experimental vaccine were eighty percent less likely to die of tuberculosis compared to children who had not received the vaccine. However, the children selected to receive the vaccine were chosen because parental consent could be obtained easily, and they may have differed systematically from children in families that refused to give consent. A follow-up study that was restricted to children who had parental consent found no difference between the death rates for vaccinated and unvaccinated children.

3 It would have been better if the experiment had been double-blind in that neither Lind nor the patients knew what they were getting. Otherwise, Lind may have seen what he wanted to see and the patients may have told Lind what they thought he wanted to hear. In a modern setting, some patients can be given vitamin C tablets, while others are given a placebo that looks and tastes like the vitamin C but is, in fact, an inert substance.

In one experiment in a senior-level course in experimental psychology, each of twelve students was given five rats to test on a maze. Six students were told that their rats were "maze-bright" because they had been bred from rats that did very well in maze tests; the other six students were told that their rats were "maze-dull." In fact, there had been no special breeding. The students were given randomly selected ordinary rats. Nonetheless, when the students tested the rats in maze runs, the rats that had been called maze-bright where given higher scores and their scores increased more over time—indicating that they were smarter and learning faster than the dull rats, but really reflecting the biased expectations of the researchers.

4 There should have been more patients. Two patients taking a medication is anecdotal, not compelling evidence. With dozens or hundreds of randomly separated patients, we can calculate the chances that the observed differences are due simply to the luck of the draw.

If two of the twelve patients studied by Lind recovered, the chances that, by luck alone, they would have received the same medication is nine percent, which is suggestive, but not compelling, evidence supporting the efficacy of the cure.

The Gold Standard

The gold standard for medical tests is a randomized controlled trial (RCT) that satisfies the four ways in which Lind's experiment could have been improved.

I *Controlled*: In addition to the group receiving the treatment, a control group receives a placebo, so that we can compare treatment to no treatment without worrying about the placebo effect or the body's natural ability to heal.

2 *Randomized*: The subjects are randomly assigned to the treatment group and the control group, so that we don't need to worry about whether the people who receive the treatment are systematically different from those who don't. In a large enough sample, differences among the subjects will average out.

3 *Blinded*: The test is double-blind so that the subjects and researchers are not influenced by knowledge of who is getting the treatment and who is not.

4 *Large*: There are enough data to draw meaningful conclusions; for relatively rare diseases, this may require thousands of patients.

The first two conditions are what gives the RCT its name; the other two conditions make the tests more persuasive.

When a study is finished, the statistical issue is the probability that, by chance alone, the difference between the two groups would be as large as that actually observed. Most researchers consider a probability less than 0.05 to be "statistically significant." Differences between the treatment and control groups are considered statistically persuasive if they have less than a one in twenty chance of occurring by luck alone.

RCTs can (and should) be used outside of medical research and are not restricted to a single "treatment." A study of different ways of teaching math might consider three new approaches in addition to the current approach. A study of fuel additives might consider four possibilities, in addition to fuel with no additive.

The wonderful thing about RCTs is that they can demonstrate cause-and-effect relationships. If patients given vitamin C are cured of scurvy, while patients given a placebo are not cured, then vitamin C is evidently responsible for the disappearance of scurvy.

This example also illustrates two other points about causation. First, causation can be a tendency, not a certainty. It is valid to say that vitamin C reduces the chances of developing scurvy and/or increases the chances of recovering from scurvy without saying that it works 100 percent of the time. The statement "chronic alcohol consumption can cause liver disease" does not mean that everyone who drinks alcohol develops liver disease, only that people who consume large amounts of alcohol are more likely to develop liver disease.

Second, it can be useful to say that "this causes that" without knowing exactly how it does so. Lind did not need to know why oranges and lemons combatted scurvy in order to prescribe them. Today, we understand how vitamin C works its magic, but there are many other cases in which we can be confident that A causes B without knowing precisely how.

A/B Tests

In our Internet-dominated world, there are lots of tricky questions about the efficacy of various "treatments." Will a website sell more gadgets if it uses a different background color? Will a newsletter get more subscribers if it displays endorsements on its main web page? Will an online club get more members if it displays a picture of the club officers?

These questions and many more can be answered with RCTs, which are commonly called *A/B tests* when used in Internet marketing. The *A/B* label refers to the fact that *A* is compared to *B*, though there could also be comparisons to *C*, *D*, and *E*.

Consider the question of background color. Perhaps a company has been using a white background on its main page and is considering switching to a sky-blue background. An A/B test would follow the protocol of all RCTs:

1 *Controlled*: A page with a sky-blue background is the treatment, and a page with a white background is the control.
2 *Randomized*: When a user types in the site's URL, a random event generator is used to determine whether the user is sent to the sky-blue page or the white page.
3 *Blinded*: The researchers do not directly monitor who is sent to which page and users do not know that they are part of an experiment.
4 *Large*: Based on past sales data, the experiment is set up to run until there are likely to have been enough sales to make a meaningful comparison between the two background colors.

After the data are collected, a statistical test is used to determine whether the observed difference in sales can be reasonably explained by the randomness involved in sending people to different pages or is, instead, evidence that the background color does matter.

Like other RCTs, A/B tests can be used to demonstrate cause and effect. Suppose that, holding everything else constant, the new background color generated far more sales than did the original background color. The most compelling explanation is that the change in the background color caused the increase in sales. As with all RCTs, causation can be a tendency, not a certainty. The site didn't go from zero percent sales to 100 percent sales, but the chances of making a sale increased. And, again, we almost certainly don't know precisely why the new background color resonated more with customers—nor do we need to know—but we can conclude that the color change caused the increase in sales.

A/B testing on the Internet has become so commonplace that a nerd joke in Internet marketing is that *A/B testing* is an acronym for "Always Be Testing."

Post Hoc **Reasoning**

When one event precedes another, there is a natural inclination to think that the first event is a good predictor of the second event. This need not be true. Many years ago, an advertisement for Club Med vacation resorts made this tongue-in-cheek observation:

"I was sitting near the water after a busy day of tennis and windsurfing," said a Philadelphia stockbroker vacationing here, "listening to classical music with a bunch of other people—it was Handel's 'Water Music'—when I looked out at the ocean. I couldn't believe my eyes.

"The tide, which had been moving in for hours, actually stopped in its tracks, turned around and started rolling out…"

The Latin phrase, *post hoc ergo propter hoc* ("after this; therefore because of this"), describes the logical fallacy of assuming that because one event follows another, it must have been caused by the first event. The playing of classical music preceded the changing of the tide, but did not cause it.

The *post hoc* fallacy is a particularly pernicious error because, unlike contemporaneous correlations, the order of events, A before B, strongly suggests causation.

Clive Granger, a Nobel Laureate in Economics, is best known for a causality test he proposed when it is not possible to do an RCT. Is there a cause-and-effect relationship between interest rates and stock prices? Between the unemployment rate and the outcomes of presidential elections? We are all thankful that economists cannot manipulate interest rates and the unemployment rate in order to collect data for their research.

Granger proposed, instead, that we look at observational data and see if one variable is helpful in predicting future values of another. If interest rates are helpful in predicting future stock prices, but not vice versa, then changes in interest rates are said to cause changes in stock prices. Such a conclusion is clearly a *post hoc* fallacy so, instead of *causality*, this test is commonly called *Granger causality*. Thus, interest rates Granger-cause stock prices.

Roosters begin crowing before dawn, announcing their territorial claims. A Granger test would observe that crowing starts before the sky becomes light, but the sky does not become light before roosters crow. Therefore, the test would conclude that crowing causes the sky to light up, and the rising sun does not cause crowing. The truth, of course, is that neither causes the other in any meaningful sense of the word. Roosters crow in anticipation of the sun rising due to a circadian rhythm controlled by an internal clock. Roosters kept in dim light around the clock crow (approximately) every twenty-four hours, when they expect the sun to rise.

It is hard to establish true causality without an RCT.

Good to So-So

RCTs are not always possible. If we want to know why some companies' stock returns are better than others, we can't take over companies and fiddle with the ways they are run and see what happens. Instead, we have to make do with observational data—we observe different companies doing different things and try to draw conclusions. This is a worthwhile goal, but there are good reasons for caution. Remember the sequence in which RCTs are run: choose the treatment and the control groups, collect data, and compare results. It can be misleading to do the reverse: look for patterns in observational data and then treat the discovered patterns as if they had come from an RCT. This is known as HARKing: Hypothesizing After the Results are Known. The harsh sound of the word reflects the dangers of HARKing.

A classic example is a 2001 study known as *Good to Great*. Gary has written extensively about this study, but we retell the story here because it is such a clear example of HARKing. After scrutinizing the stock prices of 1,435 companies over the years 1965–2000, Jim Collins identified eleven companies that had trounced the overall market:

Abbott Laboratories	Kimberly-Clark	Pitney Bowes
Circuit City	Kroger	Walgreens
Fannie Mae	Nucor	Wells Fargo
Gillette	Philip Morris	

A portfolio of these eleven stocks would earned a 19.2 percent annual return between 1965 and 2000, compared to a 12.2 percent return for the market as a whole. After identifying these eleven stocks, Collins looked for common characteristics and reported five common themes he found:

1. Level 5 Leadership: Leaders who are personally humble, but professionally driven to make a company great.
2. First Who, Then What: Hiring the right people is more important than having a good business plan.
3. Confront the Brutal Facts: Good decisions take into account all the facts.
4. Hedgehog Concept: It is better to be a master of one trade than a jack of all trades.
5. Build Your Company's Vision: Adapt operating practices and strategies, but do not abandon the company's core values.

The five characteristics are plausible and the names are memorable, but it is a lot easier to make observations about the past than to make predictions about the future.

The evidence for these five characteristics would have been persuasive if Collins had identified companies *in advance* that do and do not have these characteristics (the treatment and control groups), and then monitored their success. Instead, he peeked at the results and found five patterns. He had HARKed and he was proud of it:

> It is important to understand that we developed all of the concepts in this book by making empirical deductions directly from the data. We did not begin this project with a theory to test or prove. We sought to build a theory from the ground up, derived directly from the evidence.

It is not surprising that Collins' eleven stocks have not done as well after they were selected as they had done before they were selected. Figure 2.2 shows the performance of a portfolio of his eleven stocks compared to the overall market, starting with a $1 investment on January 1, 2002, shortly after *Good to Great* was published. The annual rate of return has been 4.7 percent for the good-to-great portfolio and 6.9 percent for the market portfolio.

This HARKing problem is endemic in formulas/secrets/recipes for becoming wealthy, having a lasting marriage, living to be 100, and so on and so forth, that are based on backward-looking studies of wealthy people, durable marriages, and long lives.

Figure 2.2 From great to so-so.

In 2019, for example, *Business Insider* reported: "if you possess a certain set of characteristics, you may be more likely to become wealthy, according to Sarah Stanley Fallaw, director of research for the Affluent Market Institute." Ms. Fallaw surveyed more than 600 American millionaires, and identified six characteristics she calls "wealth factors" that are correlated with wealth; for example, frugality and confidence in financial management. Except for having six characteristics, instead of five, this is exactly like Good to Great, and just as useless.

If we think that we know some secrets for success, a valid way to test our theory would be to identify people with these traits and see how they do over the next ten, twenty, or fifty years. Otherwise, we are just HARKing: scrutinizing the past instead of predicting the future.

Scuttlebutt

Here's an example of a study that went in the other direction—from theory to data, instead of vice versa.

Philip Fisher, a legendary investor, touted the value of "scuttlebutt" in his classic 1958 book, *Common Stocks and Uncommon Profits*. Scuttlebutt is collected by talking to a company's managers, employees, customers, and suppliers, and to knowledgeable people in the company's industry, in order to identify able companies with good growth prospects. Perhaps Wall Street is too focused on numbers (sales, revenue, profits) that fluctuate wildly quarter to quarter and is not paying enough attention to the underlying qualities that make a company successful.

It is hard to measure scuttlebutt, but Gary thought up an interesting gauge. Since 1983, *Fortune* magazine has published an annual list of the most-admired companies based on surveys of thousands of executives, directors, and security analysts. The top ten (in order) in 2020 were: Apple, Amazon, Microsoft, Walt Disney, Berkshire Hathaway, Starbucks, Alphabet (Google), JPMorgan Chase, Costco, and Salesforce.

In 2006 Gary and a student looked at the performance of a stock portfolio of Fortune's top-ten companies from 1983 through 2004. The most-admired portfolio started by investing an equal amount in each of 1983's ten most-admired companies, with the investment made on the magazine's official publication date, which is few days after the magazine goes on sale. When the 1984 most-admired list was published, the 1983 stocks were sold and the proceeds were invested in the 1984 top ten, and so on until the final investment in the 2004 ten most-admired companies.

The most-admired strategy beat the S&P 500 by 2.2 percentage points a year, with respective annual returns of 15.4 percent versus 13.2 percent. Compounded over twenty-two years, every dollar initially invested in the most-admired portfolio would have grown to $23.29, compared to $15.18 for a dollar initially invested in the S&P 500.

It is unlikely that this difference is some sort of risk premium since the companies selected as America's most admired are large and financially sound and their stocks are likely to be viewed by investors as very safe. By the usual statistical measures, they were safer. Nor is the difference in returns due to the extraordinary performance of a few companies. Nearly sixty percent of the most-admired stocks beat the S&P 500.

Perhaps Fisher was right. The way to beat the market is to focus on scuttlebutt—intangibles that don't show up in a company's balance sheet—and the Fortune survey is the ultimate scuttlebutt.

Gary revisited this strategy recently and found that the most-admired strategy has continued to beat the S&P 500. In fact, the margin for the subsequent fourteen-year period 2005–2018 was slightly better than for

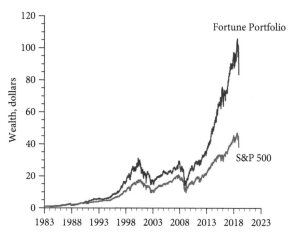

Figure 2.3 Most-admired portfolio versus S&P 500, 1983 through 2018.

Table 2.1 *Annual returns from purchases made after* **Fortune's** *cover date, 1983–2018.*

Days After Publication	Most-Admired Portfolio	S&P 500
0	13.30%	10.83%
5	13.53%	10.76%
10	13.64%	10.83%
15	13.91%	11.09%
20	13.68%	10.87%

the initial period—10.09 percent versus 7.27 percent, a 2.82 percentage point difference.

Over the entire thirty-six-year period, the most-admired strategy had a 13.30 percent annual return, compared to 10.83 percent for the S&P 500. Figure 2.3 shows that every dollar initially invested in the most-admired portfolio in 1983 would have grown to $89.48, compared to $40.58 for the S&P 500.

Table 2.1 shows that this superior performance did not depend on the annual purchases being made on the publication date. Making the purchases five, ten, fifteen, or even twenty days after the publication date would still have been a profitable strategy.

What distinguished this study of stock returns from the Good-to-Great study is that it was motivated by a plausible theory that was conceived before looking at the data that were used to test the theory—and it was then replicated with additional data. Research has a much better chance of being useful when theory comes before data, rather than the other way around.

If You Love Your Job

A student once asked Gary if he should take a job in management consulting or investment banking. Consulting has a reputation as being less exhausting, but banking is potentially far more lucrative. Gary answered with a proverb often attributed to Confucius, "If you love your job, you will never work a day of your life." An ideal job is one you wake up in the morning eager to get started, a job you never want to retire from, one you would be (almost) willing to do for free. Too many people take jobs they hate, counting the days until they retire. This is no way to live a life.

A corollary is that people who love their jobs are likely to be more productive and that a company staffed by people who love their jobs is likely to be more successful. Maybe, like the most-admired companies, identifying firms whose employees love their jobs is a valuable bit of scuttlebutt.

To test this theory, Gary and two of his students, Sarah Landau and Owen Rosebeck, looked at Glassdoor's annual list of the fifty "Best Places to Work." Glassdoor's website, launched in 2008, allows current and former employees to rate their companies on a five-point scale from 1 (least attractive) to 5 (most attractive). It has now accumulated more than fifty million reviews for nearly a million companies.

Glassdoor averages the ratings each December (with recent ratings weighted heavier than distant ratings) and publicly announces the "Best Places to Work." The top fifty companies were reported each year from 2009 through 2017; the top 100 were reported in 2018, but, for consistency, Gary's team only considered the top fifty that year.

Roughly forty percent of the top companies each year are private firms or subsidiaries of a larger company; for example, in the 2018 rankings, the top vote getter was Facebook; the next three were private companies (Bain, Boston Consulting, and In-N-Out Burger), followed by Google at number five.

It is clearly not a perfect system. All reviews are screened by Glassdoor personnel (and roughly twenty percent are rejected), but there is no practical way of ensuring that the reviews are honest or reflective of the views of other employees. People who feel strongly about the company they work for are more likely to take the time to write reviews (and more prone to hyperbole) than are people who are largely indifferent.

Still, there might be some value in the ratings. To test this theory, Gary, Sarah, and Owen compared the ten-year performance of a Best-Places-to-Work Portfolio with the S&P 500. At the beginning of each year, the Best-Places portfolio invested an equal amount in the publicly traded Best-Places companies that had been announced in December, and these stocks were held until the beginning of the next year.

Figure 2.4 shows that an initial $1 investment in the Best-Places portfolio on January 1, 2009, would have grown to $5.52 on December 31, 2018, a nineteen percent annual rate of return, while a $1 investment in the S&P 500 would have grown to $3.42, a thirteen percent annual return.

Maybe employee ratings are useful scuttlebutt.

Figure 2.4 Happy workers, happy stockholders?

Clever Tickers

Here's a more provocative study that was also motivated by theory.

Corporate stocks have traditionally been identified by ticker symbols (so-called because trading used to be reported on ticker tape machines). Companies choose their ticker symbols and have traditionally chosen abbreviations of the company's name; for example, AAPL for Apple and GOOG for Google. Sometimes, the ticker symbols become so familiar that companies become known by their tickers. International Business Machines (ticker: IBM) is now universally called IBM. Minnesota Mining and Manufacturing (ticker: MMM) changed its legal name to 3M in 2002.

During the past few decades, dozens of companies have shunned the traditional name-abbreviation convention and chosen ticker symbols that are related to what the company does. Some are memorable for their cheeky cleverness; for example, Southwest Airlines' choice of LUV as a ticker symbol was related to the fact that its headquarters are at Love Field in Dallas, and Southwest wanted to brand itself as an airline "built on love."

Those who believe that the stock market is "efficient," with a stock's price accurately reflecting its true value, dismiss the idea that stock prices might be affected by superficial things like ticker symbols. However, investors are not always as rational as efficient-market enthusiasts assume—and that includes the influence of ticker symbols. For example, there have been several notable instances where investors mistakenly bought or sold the wrong stock because they were confused about the stock's ticker symbol.

While he was at Harvard working on his economics thesis, Michael Rashes noticed that a takeover offer for MCI Communications sparked a jump in the price of a stock with the ticker symbol MCI. Unfortunately, the ticker symbol for MCI Communications was MCIC! Investors had mistakenly rushed to buy stock in MassMutual Corporate Investors, a completely unrelated company in a completely different industry because it had the ticker symbol MCI. Rashes collected several other examples and published a paper in the *Journal of Finance* with the wonderful title, "Massively Confused Investors Making Conspicuously Ignorant Choices: (MCI-MCIC)."

Investment decisions are sometimes distorted by mistakes and flawed judgments. We are only human, after all.

There is considerable evidence that human judgments are shaped by how easily information is processed and remembered. For example, statements like "Osorno is in Chile" are more likely to be judged true if written in colors that are easier to read, and aphorisms that rhyme are more likely to be judged true; for example, "Woes unite foes" versus "Woes unite enemies."

It has also been demonstrated repeatedly that experiences that elicit positive emotional arousal are more likely to be remembered and that people are more likely to have positive feelings about things that are associated with the positive experiences.

These arguments suggest that ticker symbols that are pronounceable and clever might be more easily recalled and rated favorably, which might have an effect on stock prices. We can't do an A/B test, assigning clever and dull ticker symbols to randomly selected companies and seeing whether the stock returns differ, but we can identify such companies *before* looking at their stock performance.

In 2006 Gary and two students sifted through 33,000 ticker symbols for past and present companies, looking for ticker symbols that might be considered noteworthy. Ninety-three percent of the selections coincided. They merged the lists and discarded tickers that were simply an abbreviation of the company's name (for example, BEAR for Bear Automotive Service Equipment) and kept tickers that were intentionally clever (GRRR for Lion Country Safari parks and MOO for United Stockyards).

They distributed 100 surveys with the culled list of 358 ticker symbols, the company names, a brief description of each company's business, and the following instructions:

Stocks are traded using ticker symbols. Some are simply the company's name (GM, IBM); some are recognizable abbreviations of the company's name (MSFT for Microsoft, CSCO for Cisco); and some are unpronounceable abbreviations (BZH for Beazer Homes, PXG for Phoenix Footwear Group). Some companies choose symbols that are cleverly related to the company's business; for example, a company making soccer equipment might choose GOAL; an Internet dating service might choose LOVE.

From the attached list of ticker symbols, please select 25 that are the cleverest, cutest, and most memorable.

Seasoned investment professionals were intentionally excluded from the list of people who were surveyed, as their choices might have been influenced by their knowledge of the investment performance of the companies on the list.

For each trading day from the beginning of 1984 (when clever ticker symbols started becoming popular) to the end of 2005, they calculated the daily return for a portfolio of the eighty-two clever-ticker stocks that received the most votes in the survey. As a control group, they used the overall performance of the stock market.

Figure 2.5 shows that the clever-ticker portfolio lagged behind the market portfolio slightly until 1993, and then spurted ahead. Overall, the compounded annual returns were 23.5 percent for the clever-ticker portfolio and 12.0 percent for the market portfolio. Because of the power of compound interest over this twenty-two-year period, $1 invested in the market would have grown to $12.17, while $1 invested in the clever-ticker portfolio would have grown to $104.69.

The market-beating performance was not because the clever-ticker stocks were concentrated in a single industry. The eighty-two clever-ticker companies spanned thirty-one of the eighty-one industry categories used by the U.S. government, with the highest concentration being eight companies in eating and drinking establishments, of which four beat the market and four did not. Nor was the clever-ticker portfolio's success due to

Figure 2.5 Clever-ticker portfolio performance, 1984 through 2005.

the extraordinary performance of a small number of clever-ticker stocks: sixty-five percent of the clever-ticker stocks beat the market.

Although Gary and his students had tried to exclude industry professionals who might be familiar with the clever-ticker stocks and their performance, they may have inadvertently included some people who knew how some of these stocks had done during the period being studied.

Seeking even stronger evidence, in 2019 Gary and two more students revisited the performance of this clever-ticker portfolio. The people surveyed in 2006 who chose the eighty-two clever-ticker stocks may have known something about how the stocks had performed before 2006, but they could not possibly know how these stocks would do after the survey was done.

As was true for the original twenty-two years, 1984–2005, the clever-ticker portfolio outperformed the market portfolio by a substantial margin for the subsequent thirteen years, 2006–2018. Figure 2.6 shows that, starting with $1 on the first trading day in 2006, the clever-ticker portfolio grew to $5.03, a 13.2 percent compound annual return, while the market portfolio grew to $2.47 at the end of 2018, a 7.2 percent compounded annual return.

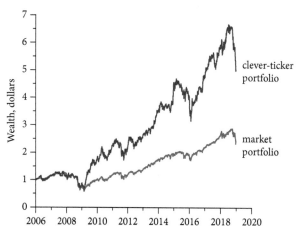

Figure 2.6 Clever-ticker portfolio performance, 2006 through 2017.

In a weird coincidence, one of Gary's former students, Michael Solomon, contacted Gary after reading the original clever-ticker article. In 2000, Michael was working for Leonard Green, a private equity investment firm, when it acquired VCA Antech, a company that operates a network of animal hospitals and diagnostic laboratories. Leonard Green reorganized VCA Antech and made a public stock offering in 2001. Michael suggested the ticker symbol WOOF and they went with it.

An investor considering pet-related companies might come across VCA Antech and barely notice the ticker symbol if it were something boring and unpronounceable, like VCAA. But the actual ticker symbol, WOOF, is memorable and funny. Perhaps a few days, weeks, or months later, this investor might consider investing in a pet-related company and remember the symbol WOOF.

When the stock went public in 2001 with the ticker symbol WOOF, some financial experts were amused and skeptical. A MarketWatch column was headlined, "Veterinary IPO barking in market." A Dow Jones Newswire story said that, "The initial public offering of VCA Antech Inc., whose stock symbol is WOOF, was—it must be said—a dog of a deal." Hey, any publicity is good publicity, right?

Figure 2.7 shows that VCA beat the market handily over the next sixteen years until it was acquired by Mars, the candy company. That

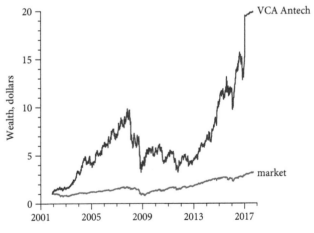

Figure 2.7 Woof! Woof!

acquisition is the reason for the price spike in 2017. Over the more than sixteen years that VCA Antech traded under the ticker symbol WOOF, the annual return on its stock was 19.4 percent, compared to 7.2 percent for the market as a whole.

There was no RCT, so we don't know for certain that the ticker symbol was responsible for the superior performance of VCA Antech and the other clever-ticker stocks, but the evidence is pretty strong. The idea makes sense and, unlike *Good to Great*, the idea was conceived and the stocks were selected before looking at their performance—a protocol reinforced by the fact that the performance was revisited a dozen years after the original study and the clever-ticker stocks were still outperforming the market.

How to Avoid Being Misled by Phantom Patterns

Coincidental correlations are useless for making predictions. In order to predict something, it has to be predictable due to an underlying causal structure. There must be a real reason for the correlation: A causes B; B causes A; A causes B and B causes A; or something else causes A and B. Correlations without causation mean predictions without hope.

Causation can be established by an RCT—a trial in which there is both a treatment group and a control group and the subjects are randomly assigned to the two groups. In addition, the test should be double-blind so that neither the subjects nor researchers know who is in the treatment group. Finally, there should also be enough data to draw meaningful conclusions. True RCTs can generally be trusted since they are motivated by plausible theories and tested rigorously.

Often, we cannot do RCTs. We have to make do with observational data. A valid study specifies the theory to be tested before looking at the data. Finding a pattern after looking at the data is treacherous, and likely to end badly—with a worthless and temporary coincidental correlation.

CHAPTER 3

Duped and Deceived

For centuries, residents of New Hebrides believed that body lice made a person healthy. This folk wisdom was based on a pattern: healthy people often had lice and unhealthy people usually did not. However, it turned out that it was not the absence of lice that made people unhealthy, but the fact that unhealthy people often had fevers that drove the lice away.

We have been hard-wired to notice (indeed, to actively seek) patterns that can be used to identify healthy food, warn us of danger, and guide us to good decisions. Unfortunately, the patterns we find are often misleading illusions. Throughout history, humans have been duped and deceived by phantom patterns.

Malaria

Malaria has been around for thousands of years. Quintus Serenus Sammonicus, a celebrated Roman physician in the second century CE, told malaria patients to wear a piece of paper with the triangular inscription shown in Figure 3.1 around their necks. The top line spells the mystical word *abracadabra*. Each succeeding line removes the last letter of the preceding line and slides the letters to the right, resulting in a repetition of each letter moving down and to the right, and the complete word *abracadabra* running from the last line up and to the right. Malaria patients were advised to wear this amulet for nine days and then throw it over their shoulder into a stream that flowed to the east.

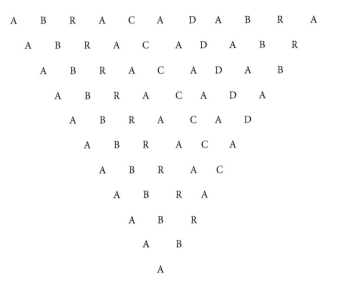

Figure 3.1 Abracadabra.

It is not surprising that some patients found comfort in the strong pattern formed by the abracadabra letters. It is unlikely that they would have had as much faith in a random collection of letters.

Some patients no doubt recovered after this nonsense and their recovery was attributed, *post hoc*, to the physician's prescription. Don't laugh. Even today, some patients take ineffective medications, recover, and ascribe their recovery to the worthless medicines.

In the summer of 1969, Gary's brother told him that a friend was making a fair amount of money selling "energy rocks" to passersby near Pier 39 in San Francisco. These stones were ordinary gravel that he had scooped up in a nearby parking lot and washed with tap water. He was there almost every day that summer, and no one ever came back to complain that their energy rocks weren't working.

In 400 BCE, Hippocrates noticed that people afflicted with what we now know to be malaria often lived in marshy, swampy areas. He described malaria as "marsh fevers" that were caused by a *miasma* "pollution" created by rotting organic matter. Miasmas were also described as "bad air" or "night air" because the smells were particularly offensive at night. The word *malaria* is thought to come from the Italian mal'aria (bad air), though many other diseases, including cholera, were also believed to be caused by miasmas. The miasma theory was based on misleading patterns.

In the case of malaria, mosquitos love swampy areas and come out at night, and it was eventually proven that it was not the foul air, but bites from *anopheles* mosquitos that transmitted the disease.

Sir Ronald Ross was a British medical doctor born in India in 1857, and was the son of a general in the British Indian Army. He was inspired by Sir Patrick Manson, considered the founder of the field of tropical medicine, to investigate whether the malarial parasite was transmitted by mosquitos. In 1897, while stationed in Secunderabad, India, he harvested several mosquitos from larvae in a laboratory to ensure that they had not been contaminated by the outside world. He then paid a malaria patient eight annas for allowing himself to be bitten by eight mosquitos. When Ross dissected the mosquitos, he found malarial parasites growing in their stomachs, which was compelling evidence that mosquitos acquired the parasite when they bit infected humans. (For a randomized controlled trial, he should have also examined a control group of mosquitos that had not been allowed to bite the malarial patient, in order to rule out the possibility that malarial parasites were passed from mosquito parents to offspring.)

He celebrated his discovery by writing a poem that included these lines:

> With tears and toiling breath,
> I find thy cunning seeds,
> O million-murdering Death.
> I know this little thing
> A myriad men will save.
> O Death, where is thy sting?
> Thy victory, O Grave?

In later experiments with birds, Ross demonstrated that *anopheles* mosquitos transmit the parasite by biting infected birds and then biting healthy ones, thereby completing the cycle by which malaria spreads.

In 1902, Ross became the first British Nobel Laureate when he received the Nobel Prize for Physiology or Medicine for his malarial work.

Superstitions

A central part of the eighteenth century Age of Enlightenment was a celebration of the scientific method—an insistence that beliefs are not to be accepted uncritically, but be tested empirically. Baseless superstitions do not fare well when tested scientifically. Nonetheless, many superstitions have been hard to shake, even if they don't make sense and we have nothing more than hunches about their origins. Perhaps it is considered bad luck for a black cat to cross your path because black cats are associated with witches. Knocking on wood for good luck may have come from Christians touching wood as a reminder of the cross that Jesus was crucified on. Bad luck from opening an umbrella indoors probably came from the ancient Egyptians who had sun umbrellas and thought that the sun god, Ra, would be offended if an umbrella were opened indoors where it was not needed. Who knows why wishing on a star (especially a shooting star) or finding a four-leaf clover is considered good luck?

No doubt, many people recite superstitions for amusement, but don't really believe them. Others may believe them, but their convictions are relatively harmless. Either way, most superstitions are difficult to test empirically. How do we set up treatment and control groups? If we knock on wood and wait for good luck, how long do we wait? What counts as good luck?

No doubt, selective recall and confirmation bias reinforce the allure of superstitions. We tend to remember the hits that validate our beliefs and forget the misses that contradict them. If a novice wins a game, we chant "beginner's luck" because it confirms our belief in beginner's luck. If a novice loses, we invite them to play again (and forget this evidence against beginner's luck). If someone walks under a ladder and something bad happens a few hours, days, or weeks later, we might remember the ladder. If nothing bad happens, we forget the ladder.

Imagining Patterns

Our addiction to patterns is so strong that we sometimes imagine them. Gary has a relative (Jim) who is nuts about the Boston Celtics, but Jim stopped watching their games on television after he noticed a few games when the Celtics did poorly while he was watching. He jinxes the Celtics! Logically, the Celtics players do not know or care whether Jim is watching, so there is no way that their play could be affected by his viewing habits. Yet he believes.

We don't know (or want to know) what psychological factors are responsible for people thinking that they are walking curses, so that everything they touch turns to dross. We are pretty sure, though, that selective recall is a large part of the story. Some people are more likely to remember the good, others are more likely to remember the bad. There were surely times when Jim watched the Celtics play well, but what stuck in Jim's memory were those time when the Celtics did poorly while he was watching.

Gary once suggested a controlled experiment. Pick a game and then randomly select time intervals to watch and not watch, and see whether there is a difference. Jim refused to do it. Perhaps this experiment was too nerdy, or perhaps Jim didn't want the power of his curse to be challenged.

Memorable M Patterns

On the eve of the 1992 U.S. presidential election, the Minnesota Vikings played the Chicago Bears on Monday Night Football. At the beginning of the television broadcast, the announcers revealed the Vikings Indicator: when the Minnesota Vikings win a game on the Monday before a presidential election, the Republican candidate *always* wins the election. It turns out that this indicator was based on exactly two observations: a 1976 Vikings loss before Democrat Jimmy Carter was elected, and a 1988 Vikings win before Republican George Bush was elected. The third time was not a charm: Minnesota won the 1992 game and George Bush lost to Democrat Bill Clinton.

One of Gary's students, Gunnar Knapp, discovered several other M patterns. Montana Senator Mike Mansfield was from Missoula, Montana, was born in March, was a marine, married a woman named Maureen, attended the Montana School of Mines, and was the longest serving Majority Leader of the Senate. When he retired from the Senate, he was succeeded by John Melcher, who had attended the University of Minnesota before enlisting in the military. Gunnar also found that Middlesex County, Massachusetts had a congressman named McDonald from the city of Malden. We could go on, but we won't.

You Don't Split

Jay and a friend, Rory, were once in Las Vegas and sat down at a blackjack table to chat with a friend of Rory's who had been playing at the table alone. The cards suddenly turned against Rory's friend and it wasn't long before he got up and went to another table without saying a word. There was little doubt that he was trying to escape the curse that Jay and Rory had placed on him. When Rory later went over to tell him that his wallet had fallen out of his pocket, the friend told Rory to get away from him. After Rory gave him his wallet and left, the friend waved his hands in the air wildly in an effort to fan the bad luck away as if it were a stench that Rory had left behind.

Anyone who has played games at a casino knows that superstitious beliefs are alive and well. Blackjack players often get truly annoyed if a player makes a bad decision that changes the future cards other players are dealt (for example, unwisely taking a card from the deck that would otherwise have gone to the complaining player). Those unfortunate outcomes are remembered while the positive repercussions of bad choices are forgotten.

Blackjack players with lots of experience can sometimes become very agitated when another person makes a play they wouldn't have made. Jay learned this firsthand at his first job after college graduation. He had played very few games of blackjack, but he had read Edward Thorpe's classic book *Beat the Dealer*. Jay was discussing blackjack with a co-worker named Robby and an argument broke out about a relatively obscure question—whether it is better to stand or split two nines against a dealer showing a six.

The goal in Blackjack is to get closer to twenty-one (without going over) than the dealer, who must take another card if she has a total of sixteen or lower. Being dealt two nines (a total of eighteen) when the dealer has a six showing is great for you, because face cards (which count as ten) are the most likely cards in the deck. This means that the dealer is likely to be stuck with the most unfavorable total possible (sixteen) and has a high probability of going over twenty-one. For instance, if you stand with your eighteen and she's stuck with a sixteen, you would only lose if her next card is a three, four, or five (a two would be a tie). The intuitive play is to "stand" at eighteen and wait for the unfortunate dealer to go bust with a total above twenty-one.

However, there is a complication in that if you have a pair, you are allowed to split your cards into two separate hands (and win or lose twice as much). Here, you could split your pair of nines into two hands, each with a nine, so that each hand can be dealt additional cards. If you get lucky and each hand is dealt a face card, then you have two strong hands with nineteen instead of one good hand with eighteen. Is the bird in the hand better than two in the bush?

Based on his own playing experience, Robby strongly believed that you don't let the bird out of your hand. Jay disagreed, based on what he had read, and he couldn't believe that Robby trusted his imperfect recollection of anecdotes more than well-researched analysis.

Robby was at his wit's end because no matter how eloquently and passionately he argued his case, Jay was not about to trust him more than a respected mathematician who had analyzed millions of computer simulations.

Then Jack, the company's highest-level actuary, happened to walk by the cubicle. He was in upper management and wasn't known for chatting with workers far down the chain of command, which made the exchange even more surprising:

Robby: Hey Jack, you're a smart guy, come here!

Jack: (looking confused and hesitating before walking over)

Robby: Okay, in blackjack, the dealer's showing a six and you've got two nines. What do you do?

Jack: I don't know; I don't play cards.

Robby: Well let me tell you: YOU DON'T SPLIT!!

Jay still laughs at the incongruous sight of Robby, who happened to be a bodybuilder, screaming at the much smaller and very much senior manager, who had no idea where the anger was coming from.

Be wary of people who rely on anecdotes and selective recall, especially if they might be on steroids.

If You Believe, It Will Happen

Sometimes, a superstition becomes a self-fulfilling prophecy. There are many variations of this classic story: A dance instructor convinces an aspiring dancer that her magic shoes make her a great dancer. Then she is distraught when she forgets to bring her shoes to an important recital. The instructor reveals that there is nothing magical about her shoes. Realizing that she is a good dancer, she performs brilliantly.

In real life, Pelé, considered the best footballer of all time, once had a few uncharacteristically poor games and decided that this was because a game jersey that he had given to a fan must have been a lucky jersey. Pelé told a friend to do whatever it took to find his lucky jersey and, sure enough, when Pelé's jersey was returned, his success returned, too. The friend did not tell Pelé that he had been unable to find the lucky jersey; so, he had simply given Pelé the same jersey he had been wearing during his run of bad games.

When Gary gives a final examination in his classes, students will sometimes show up for the test dressed in business attire; other students show up in their pajamas. Maybe they think their unusual clothing will bring them good luck. Maybe they have more confidence when they dress for success or wear comfortable clothing.

Confidence is undeniably important and maybe even misplaced confidence can sometimes be helpful.

Monday the 6th

Virtually every culture has lucky and unlucky superstitions that have been around for so many generations that no one knows for sure how the superstitions got started. Many of these superstitions involve lucky and unlucky numbers, even though the seemingly arbitrary ways that lucky and unlucky numbers vary from one culture to the next is ample evidence that these superstitions have no rational basis. If a number were truly unlucky, it would be unlucky around the world.

The number "13" is considered unlucky in many western cultures. Some say that "13" is unlucky because Judas Iscariot, who betrayed Jesus, was the thirteenth person to arrive for the Last Supper. Others say that there are thirteen full moons in a calendar year, and full moons make people behave strangely. Yet others say that thirteen is a bad number in comparison to twelve, which is a special number because there are twelve months of the year, twelve zodiac signs, twelve apostles, and twelve is the last number before the teens (thirteen, fourteen, and so on). Whatever the reason, the fear of the number "13" can be so overwhelming that there is a name for it: *triskaidekaphobia*.

Many tall buildings in western countries have twelfth and fourteenth floors, but no thirteenth floor, as in Figure 3.2. Jay has sometimes been tempted to go to the fourteenth floor and point out, "You know, just because they CALL this the fourteenth floor doesn't mean that it is. You have twelve floors under you, which means that you are actually on the thirteenth floor." So far, he has been too polite to do this.

Friday the thirteenth—which happens one to three times every year—is, of course, considered especially unlucky, perhaps because Jesus was crucified on a Friday (though not Friday the thirteenth). However, in

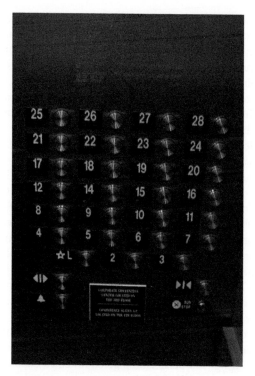

Figure 3.2 Notice anything missing?
By Sgerbic - Self-photographed, Public Domain

Italy, "13" is thought to be a lucky number, while "17" is an unlucky number, and it is Friday the *seventeenth* that is an unlucky day. In Greece and many Spanish-speaking countries, it is *Tuesday* the thirteenth, not Friday the thirteenth, that is unlucky.

Greeks are also said to consider "3" an unlucky number because, "Bad luck comes in threes," while Swedes consider "3" to be a lucky number because, "All good things must come in threes." As we said, if "3" were genuinely lucky or unlucky, it would be consistently so—not lucky in Sweden and unlucky in Greece.

In Japanese, Mandarin, and Cantonese, the pronunciation of *four* and *death* are very similar, which makes "4" an unlucky number, while the pronunciation of *eight* is similar to *wealth* or *prosper*, which makes "8" a lucky number. (By the way, Japanese actually has an alternate way of pronouncing *four* ("yon") because of the unfortunate pronunciation imported from Chinese.)

In Asia, the fourth floor is often missing. In hospitals in China and Japan, even rooms numbered "4" are missing. Many people go to a great deal of trouble to obtain lucky home addresses, telephone numbers, and automobile license plate numbers. Some of Gary's relatives refused to even consider buying an otherwise attractive home because the house number was unlucky.

Jay worked as a data analyst on the forty-fourth floor of a building in downtown Los Angeles. It turns out that when you say "44" in Chinese it's "sìshísì," which happens to sound very similar to "sǐ shì sǐ," which literally means "die, yes die". It also doesn't help that putting yes between repeated words is a common grammatical structure that equates to saying it's really true, so saying forty-four sounds like saying "must die". Needless to say, "44" is not considered a lucky number in Chinese.

One of Jay's Chinese-American co-workers did not tell her mother that she worked on the forty-fourth floor. A prospective Chinese employee turned down a job offer because of the address!

Another co-worker recounted a funny story he learned while visiting Disneyland in China. Disneyland executives were initially surprised that no one was buying their personalized green Robin Hood and Peter Pan hats. Then, they learned that in Chinese culture, a green hat means that your spouse is cheating on you. When Jay expressed surprise that pranksters didn't buy the hats for practical-joke gifts, his Chinese manager perceptively responded that "perhaps the Chinese sense of humor is different than yours."

The number 666 is widely viewed as a satanic number, with an extreme aversion to the number being labeled *hexakosioihexekontahexaphobia*. Jay once worked for a company that had to confront this dreaded number. The company ran experiments on web pages and assigned a test ID to each experiment. When some workers noticed that the number of experiments had gone past 600 and was approaching 666, a debate broke out about whether or not to skip that number and go straight from 665 to 667. Rory, the manager of the testing pipeline, laughed off the concerns despite growing pressure from colleagues who did not consider it a laughing matter. Why wouldn't he skip 666 since there was little cost and would make a lot of people feel better? Maybe he wanted to make a point, and help people confront their *hexakosioihexekontahexaphobia*. Experiment 666 came and went and no demons were summoned (as far as we know).

A pedestrian did happen to be killed by a bus on the street corner next to the building. However, it wasn't the day after experiment 666; it was the morning after a late-night email went out telling everyone to develop their "hit by a bus backup plan." Make of that what you will.

There is scant evidence that some numbers are particularly lucky or unlucky, though it can be a self-fulfilling prophecy if our fear of something causes what we fear to happen. For example, patients who have surgery on an "unlucky" day may fare poorly because they are emotionally distraught, while someone who works harder on a "lucky" day may accomplish more. There is also surely a combination of selective recall and confirmation bias. We are more likely to notice and remember when something bad happens on an "unlucky" day than when it happens on other days.

Some enterprising people sell dream guides to the gullible, promising to translate a person's dreams into winning lottery numbers. One guide says that it will help you, "Learn to unlock the power of your dreams by converting images into lucky numbers and try them on lotto or power-ball games. You will be amazed how your dreams can make you rich." For example, readers are advised to bet on the number "34" if they dream of steak and the number "10" if they dream of eggs. Who knows what they are supposed to do if they need to pick six numbers and only dream about steak and eggs.

This audacious guide is a self-published forty-four-page paperback selling for $29.99.

Since winning numbers are chosen randomly and are not affected by anyone's dreams, we are not surprised that the author chose to sell a dream guide instead of buying lottery tickets. We are also saddened by people with modest income who buy dream guides and lottery tickets.

In 1987, a year with three Friday the thirteenths, the chief economist at a Philadelphia bank reported that in the past forty years there had been six other years with three Friday the thirteenths, and a recession started in three of those years. We don't think he was joking. We *do* think he had far too much time on his hands and had been fooled by a phantom pattern. Somehow, 1987 escaped without a recession.

Sometimes, people simply notice patterns. Other times, they actively search for them. An article in the prestigious *British Medical Journal*

Table 3.1 *Numbers of admissions for South West Thames residents by type of accident.*

Cause	Friday the sixth	Friday the thirteenth
Falling	370	343
Transportation	45	65
Poisoning	37	33
Animals	1	3
Undetermined	1	4
Total	**454**	**440**

compared the number of hospital admissions in the South West Thames region of England on the six Friday the thirteenths that occurred during a four-year period with the number of admissions on the preceding Friday the sixths. They first compared emergency room admissions for accidents and poisoning on the sixth and thirteenth, and did not find anything statistically persuasive. So, they looked at all hospital admissions for accidents and poisoning, and again found nothing. Then they separated hospital admissions into the five sub-categories shown in Table 3.1: accidental falls; transportation; poisoning; injuries caused by animals and plants; and not determined whether accidental or intentional.

Overall, there were more hospital admissions on the sixth, but there was one category, transportation, where hospital admissions were higher on the thirteenth. So, they concluded their study with this dire warning: "Friday 13th is unlucky for some. The risk of hospital admission as a result of a transport accident may be increased by as much as 52%. Staying at home is recommended."

This is clear example of "Seek a pattern, and you will find one." Even though there were more hospital admissions on the sixth than the thirteenth, the researchers persisted in searching for some category, any category, until they found what they wanted to find.

"47" Everywhere

Gary teaches at Pomona College and Jay graduated from Pomona College, so we naturally have a strong affinity for the college's magical number "47". In 1964, a legendary statistics professor named Donald Bentley showed his students a whimsical geometric proof of the proposition that all numbers are equal to "47", apparently in support of a student project that was compiling a list of sightings of the number "47".

As with all lucky/unlucky numbers, a large part of the "47" story is selective recall. We are bombarded by numbers every day and we notice ones that we consider lucky or unlucky, that match our birthday, or reflect some other coincidence. For Pomona people, we are on high alert for the number "47" and let "46" and "48" pass without noticing. From the west, you can drive to Pomona College by taking exit "47" on the San Bernardino Freeway. Coming from the east, you would take exit "48", but who cares about "48"? Or you could take the Foothill Freeway and get off at exit "50" or "52", depending on whether you are coming from the west or east, but, again, who cares?

The top row of the organ in Pomona's Lyman Hall has forty-seven pipes; don't ask about other rows. Pomona Graduate Richard Chamberlain was the forty-seventh person in line to be rescued in the film *The Towering Inferno*; ignore his other films. Pomona's Mudd-Blaisdell Hall was completed in 1947 and has forty-seven letters in the dedication plaque; pay no attention to Pomona's eighty-two other buildings.

Looking outside Pomona:

- Tolstoy's novel *The Kreutzer Sonata* is named after Beethoven's *Opus 47*.
- The New Testament credits Jesus with forty-seven miracles.
- The Pythagorean Theorem is Proposition 47 of Euclid's Elements.
- Caesar proclaimed "veni, vidi, vici" in 47 BCE.
- The tropics of Cancer and Capricorn are located forty-seven degrees apart.

Pretty impressive, unless you think about it (and we hope you do). How many millions, or billions, or trillions of times have numbers between "1" and "100" appeared throughout history and in our everyday lives? There are surely a very large number of "47"s (and "46"s, "48"s, and other numbers, too). Search for any number and you will find it.

There is a *47 Society* where people report their "47" sightings. For example:

My friend Tim's hockey number is 47. Later, he told us that he started noticing the number 47 coming up a lot. At first it was just a joke, but then I started noticing it. Things like getting a score of 47 in darts, finding phone numbers with 47 in them . . .

Seek "47" and you will find it. (After editing this section, Jay noticed that his phone battery was at forty-seven percent! The next time he edited this section, he checked again, and it was forty-three percent. So close)!

In addition, as with all lucky/unlucky numbers, part of the "47" story is a self-fulfilling prophecy. Pomona students have used "47" liberally. A Pomona graduate, Joe Menosky, has been a writer and co-producer for many Star Trek episodes and sprinkles "47" (and its reverse, "74") throughout them liberally: the Enterprise was built in Sector 47, the crew stops at Sub-space Relay Station 47, there were forty-seven Klingon ships destroyed, there are forty-seven survivors on a planet, and one person is shrunk to forty-seven centimeters.

J. J. Abrams, Star Trek director and producer, picked up the baton and continued the tradition in his other productions. *Mission Impossible: Ghost Protocol* ends on Pier 47. The thermal oscillator in *Star Wars: The Force Awakens* is in Precinct 47.

The next time you notice a "47", think about whether it is selective recall or another example of a Pomona student spreading the number. Don't feel obligated to report it to the *47 Society*.

Numerology

Western numerology attributes its origins to the Greek philosopher Pythagoras, a mathematician and mystic who is credited with many mathematical discoveries, including the Pythagorean theorem we learned in school, but it seems he did not initiate what is now called numerology.

Modern numerology translates a person's full name into a mathematical number using the translation code shown in Table 3.2 that assigns numbers to the letters of the Latin alphabet. Thus, letters "a", "j", and "s" are all assigned the number "1".

Table 3.2 *The numerology code.*

1	2	3	4	5	6	7	8	9
a	b	c	d	e	f	g	h	i
j	k	l	m	n	o	p	q	r
s	t	u	v	w	x	y	z	

In order to determine a person's name number, the numerology code is used to determine the number for each letter of a person's full name and then the *sum of digits*. Using Gary as an example:

Gary: $7 + 1 + 9 + 7 = 24$
Nance: $5 + 1 + 5 + 3 + 5 = 19$
Smith: $1 + 4 + 9 + 2 + 8 = 24$
Gary Nance Smith: $24 + 19 + 24 = 67$

We then add the individual digits of the full-name number until we get a single-digit *root number*:

$$6 + 7 = 13$$
$$1 + 3 = 4$$

Gary's name number is "4". This is his *destiny number*, which is said to reveal the talents and abilities he was born with. Oddly enough, if Gary were to change his name legally, his root number would change, too, and he would presumably have different talents and abilities. Jay's birth name is James, so he presumably has a split personality.

Gary also has a birth number based on his birthdate, November 11, 1945:

$$(1 + 1) + (1 + 1) + (1 + 9 + 4 + 5) = 23$$
$$2 + 3 = 5$$

Gary's birth number is 5, which is said to be his *life path*. Oddly enough, everyone born on the same date has the same life path.

Name numbers and birth numbers are then converted into human traits: (1) leader, (2) mediator, (3) communicator, (4) teacher, (5) freedom seeker, (6) nurturer, (7) seeker, (8) ambitious, and (9) humanitarian.

Gary was evidently born to be a teacher and his life path is a freedom seeker, which is okay with him—though he would be happy with any of the other possibilities.

Various numerologists use somewhat different words for the traits. Not surprisingly, the assigned words are cryptic and ambiguous—so most anyone would find them plausible and comforting, no matter what their name or birthdate.

Gary has occasionally done an interesting experiment in his statistics classes. On the first day of class, he asks the students to fill out a survey that includes their sex, birth date, and several questions that he will use in later classes; for example, "How many hours have you slept during the past twenty-four hours?" Sometimes, he comes to the second class with a set of astrological readings based on each student's date of birth. He asks each student to consider the reading carefully and give it a grade (A to F) based on its accuracy.

The grades are overwhelmingly As and Bs, even though Gary sometimes passes out randomly determined readings and, other times, gives everyone exactly the same reading. This is yet another example of "seek and you will find."

Numerology and astrology might be thought of as cheap entertainment, but there can be real costs if people make bad decisions because of their numerological or astrological readings.

A friend told Jay about a married couple who would not make any major decisions until after they had consulted a large astrology book filled with complicated charts that revealed hourly energy levels. One day, when they went to a car dealership to buy a car, they made it clear that the sale needed to be finalized before 5 p.m., at which point the energy levels would change for the worse. (Would they turn into pumpkins?) The salesman reassured them that everything would be completed before 5 p.m.

They were minutes away from signing the final papers when the clock hit 5 p.m. To the salesman's shock, the couple apologized and told him that they would have to come back the next day to finish the deal. They left the dealership without the car and returned the next day, when the energy levels had turned positive again. If they valued their time, there were some substantial costs to this superstition.

Our innate desire for order in our lives predisposes us to look favorably on analyses and advice that are based on astrological readings, destiny numbers, and other patterns or pseudo-patterns that help us find comfort in the face of so much uncertainty about the world and ourselves.

In the scientific world, numerology is held in such low regard that some cynical scientists dismiss far-fetched patterns discovered by their colleagues as "numerology."

Cosmic Coincidences

Many people have spent countless hours discovering peculiar coincidences in the virtually endless stream of numbers that measure various aspects of the universe. For example:

- radius of the Moon = 1,080 miles = 3(360) = 3(1/2)(1)(2)(3)(4)(5)(6).
- radius of the Earth = 3,960 miles = 11(360) = 11(1/2)(1)(2)(3)(4)(5)(6).
- radius of Moon + radius of Earth = 5,040 miles = (1)(2)(3)(4)(5)(6)(7).
- diameter of earth + diameter of moon = 2(3,960) + 2(1,080) = 10,080, which is the number of minutes in a week.

These are all striking; however, the moon has an equatorial radius of 1,080 miles and a polar radius of 1,079 miles, while Earth has an equatorial radius of 3,963 miles and a polar radius of 3,950 miles. Pattern seekers use the approximate numbers 1,080 and 3,960 because these are multiples of 360, which conveniently factors into (1/2)(1)(2)(3)(4)(5)(6).

Another peculiarity is that the sum of digits of the diameters of Sun, Earth, and Moon are all "9":

- diameter of the Sun is 864,000 miles $8 + 6 + 4 + 0 + 0 + 0 = 18$ and $1 + 8 = 9$.
- diameter of the Earth is 7,920 miles $7 + 9 + 2 + 0 = 18$ and $1 + 8 = 9$.
- diameter of the Moon is 2,160 miles $2 + 1 + 6 + 0 = 9$.

The number "9" is special because it appears in many spiritual and mystical contexts; for example, the nine human traits in numerology, the nine enneagram personality traits, the nine biblical gifts of god, the number of Brahma (the Creator in Hinduism), and of course, our book, *The 9 Pitfalls of Data Science*. The number "9" also has the remarkable mathematical property that the sum of the digits of any number multiplied by nine is nine.

However, it takes a bit of bending and twisting to get the sum of the digits of the diameters of the Sun, Earth, and Moon to be "9", since the actual mean diameters are 864,938 miles, 7,918 miles, and 2,159 miles, respectively.

Pattern seekers can not only ransack a virtually unlimited number of measurements, but also can use miles when that works and kilometers when that works better. For coincidences involving time, they can consider a variety of units, including years, months, weeks, days, hours, minutes, and seconds.

Energized by all the possible patterns created by this flexibility, some pattern-seekers have interpreted the mathematical curiosities that they spent long hours discovering as evidence that a god has created the universe, since a god would surely use a carefully organized master blueprint, and not put randomly sized objects in random places. Thus, one pattern collector declared that:

our job has to be to try to learn what system The Creator used and surprisingly it appears that he used a simple 9 × 11 grid and the ratios 7 and 11 and also 14.

One pattern seeker was particularly impressed by the fact that the Sun and Moon are very different sizes, but their respective distance from earth make them appear the same size:

The Creator chose to place planets at the correct distance apart so that on certain occasions we would see the amazing harmony of the master work.

This is a striking coincidence, but it is not an eternal one. In the short run, the distance from the Moon to the Earth changes continuously during its orbit due to gravitational forces. In the long run, the moon was once much closer to the earth and is now gradually moving away from the earth. As Thomas Huxley once said, this is "the slaying of a beautiful hypothesis by an ugly fact."

It's All About Us

It is easy to understand why people once believed that the sun revolves around the earth. Every day, they see with their own eyes that the sun rises in the east, moves across the sky, and sets in the west. That's about as reliable a pattern as one can hope for.

It is also consistent with our sense of self-importance to believe that we are the center of the universe with the sun, moon, and stars revolving around us. This belief was so strong that it is enshrined in the bible: "God fixed the Earth upon its foundation, not to be moved forever."

Aristarchus first proposed a Sun-centered Solar System 1,700 years before Nicolaus Copernicus wrote his book *On the Revolutions of the Heavenly Spheres*. It isn't that the sun revolves around the earth, but that the earth spins on its axis towards the east, creating an illusion of the sun moving around the earth from east to west. Aristarchus's Sun-centered model should have won converts with its simplicity and the fact that it explained the strange zig-zagging movements of planets when viewed from the moving Earth. The competing Ptolemaic Earth-centered model involved "deferents," "epicycles," "eccentrics," and "equants" that required planets to follow a complicated set of circles nested within circles. King

Alfonso X of Castile and Leon once complained: "If the Lord Almighty had consulted me before embarking upon Creation, I should have recommended something simpler."

The ironic thing about the overly complex Earth-centered model is that it actually worked! It predicted the planetary orbits better than the Sun-centered model because, at the time, orbits were assumed to be circular (they're actually slightly ellipsoidal). Those nested circles allowed the model to closely approximate the ellipsoidal planetary motions and provide reliable and accurate predictions.

It wasn't until a generation after Copernicus's book that Johannes Kepler studied the data collected by Tycho Brahe closely and determined that (1) orbits are elliptical, (2) planets move at varying speeds, and (3) the Sun is not quite at the center of these orbits. With these new facts, the Sun-centered model now made better predictions than the Earth-centered model. But it still wasn't accepted.

Galileo is now known as the father of astronomy and his famous high-powered telescope provided the finishing touches. The discovery of moons orbiting Jupiter demolished the claim that everything orbited around the earth. In addition, Galileo observed phases of Venus (variations of light on the planet's surface) that had been predicted by Copernicus's Sun-centered model.

This should have settled the debate, except that the Catholic Church was inexorably committed to the biblical assertion that the Earth is fixed on its foundation and does not move. In 1616, the belief in the Sun-centered model was declared heretical and the stage was set for the famous battle between Galileo and the Catholic Church.

Galileo had known Pope Urban VIII since his years at the University of Pisa and after a few discussions, felt he had the Pope's blessing to write a book presenting the various competing views and arguing for the superiority of the Sun-centered model. The book he produced was probably not what the Pope expected and, rather than settling the debate, threw fuel on the fire.

Galileo's *Dialogue Concerning the Two Chief World Systems* seemingly mocked the Church's position through the character Simplicio, who was presented as a man as intelligent as his name suggested. Some of Simplicio's dialogue bore a striking resemblance to the Pope's statements about astronomy, and may have added urgency to the Inquisition's demands that Galileo come to Rome and stand trial.

Galileo traveled to Rome and faced his accusers for more than two weeks. He was forced to recant his views, and was sentenced to indefinite house arrest, rather than torture. Less than ten years later, Galileo died without seeing his Sun-centered model widely accepted. The phantom pattern that humans witnessed every day of the sun rising in the east and setting in the west was just too entrenched for most people to consider a different perspective.

Bode's Law

Distances within our solar system are measured in astronomical units (AU), the average distance between the Earth and the Sun. In the eighteenth-century, two German astronomers, Johann Titius and Johann Bode, noticed the regular pattern shown in Table 3.3 in the distances of the four planets nearest to the sun,

Mathematically, this pattern can be expressed as what has come to be called the Titius–Bode law, or just Bode's law:

$$\text{Distance} = 0.4 + 0.3(2^{n-2}) \qquad \text{if n = 2 (Venus), 3 (Earth), or 4 (Mars).}$$

The Bode's law equation, distance $= 0.4 + 0.3(2^{n-2})$, isn't much of a law, since it really only gives the distances of Venus and Mars relative to Earth, and doesn't apply to Mercury unless n is arbitrarily set equal to minus infinity.

However, Table 3.4 and Figure 3.3 show that the law would also work pretty well for Jupiter and Saturn, the other two known planets at the time, if there were a planet between Mars and Jupiter.

This was a pattern with no underlying reason, but it confirmed the beliefs of many that God had arranged the planets according in a deliberate

Table 3.3 *Distances of the four planets nearest to the Sun.*

Order n	Planet	Distance (AU)	Pattern
1	Mercury	0.39	0.4
2	Venus	0.72	0.4 + 0.3
3	Earth	1.00	$0.4 + 0.3(2)$
4	Mars	1.52	$0.4 + 0.3(2^2)$

Table 3.4 *The six known planets when Bode's law was conceived.*

Order n	Planet	Distance (AU)	$D=0.4+0.3(2^{n-2})$
I	Mercury	0.39	0.55
2	Venus	0.72	0.70
3	Earth	1.00	1.00
4	Mars	1.52	1.60
5	?		2.80
6	Jupiter	5.20	5.20
7	Saturn	9.55	10.00

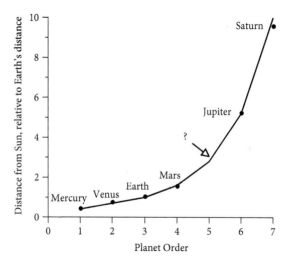

Figure 3.3 Bode's law bodes well.

pattern. Noting the gap between Mars and Jupiter, Bode asked, "Can one believe that the Founder of the universe had left this space empty? Certainly not." Both Titius and Bode encouraged astronomers to use the pattern they had discovered to search for new planets.

In 1781 Uranus was discovered reasonably close to where Bode's law predicted the next planet past Saturn would be and, in 1801, Ceres was discovered between Mars and Jupiter, close to where Bode's law predicted the fifth planet from the sun would be. Surely this was astonishing evidence of God's plan for the universe.

If you haven't heard of the planet Ceres, it is because it is no longer considered a planet. In 1930, Bode's law was further undermined by the discovery of Neptune and Pluto far from where Bode's law says they should be (Table 3.5 and Figure 3.4).

Several people tried to resuscitate Bode's law by modifying it, including S. B. Ullman, who proposed this complicated extension:

$$\text{Distance} = 0.4 \qquad\qquad\qquad\qquad \text{if } n = 1 \text{ (Mercury);}$$
$$0.4 + 0.3(2^{n-2}) \qquad\qquad \text{if } n = 2 \text{ (Venus), 3 (Earth), 4 (Mars),}$$
$$\qquad\qquad\qquad\qquad\qquad 5 \text{ (Ceres), 6 (Jupiter), 7 (Saturn), or}$$
$$\qquad\qquad\qquad\qquad\qquad 8 \text{ (Uranus);}$$
$$0.4 + 0.3(2^{n-2}) - ((n-8)3)^2 \qquad \text{if } n = 9 \text{ (Neptune) or 10 (Pluto).}$$

Such well-meaning efforts are an example of what we now call *overfitting*, a relentless addition of complexity intended solely to make a model fit the data better. Disparaging overfitting, the great mathematician John von Neumann once said, "With four parameters I can fit an elephant and with

Table 3.5 *Bode's law works for seven out of ten (?) planets.*

	Planet	Distance (AU)	$D = 0.4 + 0.3(2^{n-2})$
1	Mercury	0.39	0.55
2	Venus	0.72	0.70
3	Earth	1.00	1.00
4	Mars	1.52	1.60
5	Ceres (?)	2.77	2.80
6	Jupiter	5.20	5.20
7	Saturn	9.55	10.00
8	Uranus	19.22	19.60
9	Neptune	30.11	38.80
10	Pluto	39.54	77.20

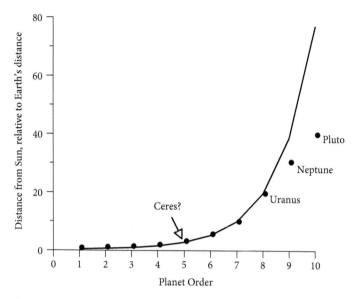

Figure 3.4 Bode's law works for the non-planet Ceres, but not for Neptune and Pluto.

five I can make him wiggle his trunk." Figure 3.5 shows a four-parameter model that does indeed look like an elephant.

The end result of overfitting is often, as with the modified Bode's law, a model that fits the data well, but has no underlying rhyme or reason—so it doesn't work well making predictions outside the limited realm in which it was manipulated and tweaked to fit the data.

In 2008, the International Astronomical Union (IAU) demoted Pluto from *planet* to *dwarf planet* and recognized four other dwarf planets: Ceres, Haumea, Makemake, and Eris. Now Bode's law had an unsolvable conundrum. If we omit the dwarf planets, like Ceres and Pluto, then Bode's law doesn't work for Jupiter, Saturn, Uranus, and Neptune. If we include the five dwarf planets, then Bode's law works for Ceres, Jupiter, Saturn, and Uranus, but doesn't work for Neptune, Pluto, and three of the dwarf planets.

What's one to do? The best response, no doubt, is to acknowledge that, over billions of years, our solar system stabilized with well-spaced planets

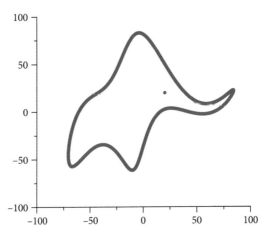

Figure 3.5 John von Neumann's elephant.

that were not disrupted by gravitational attractions among the planets, but it is a mistake to think that the spacing conforms to a mathematical one-size-fits-all "law." Humans are comforted by the idea that everything is governed by numerical patterns that are waiting to be discovered. However, patterns without reason are unreliable. Bode's law is an interesting pattern with no known uses.

Moore's Law

In 1965 Gordon Moore, the co-founder of Fairchild Semiconductor and Intel, wrote a paper provocatively titled, "Cramming more components onto integrated circuit chips." Using four years of data on the number of transistors per square inch on integrated circuits, he drew a graph like Figure 3.6. The units on the vertical axis are the natural logarithms of the number of transistors per square inch, so that the slope of the fitted line shows the rate of growth. Here, the rate of growth was roughly 1, or 100 percent, which means a doubling every year.

This rapid increase in the number of transistors has come to be known as Moore's law, even though Moore did not contend that it was a physical law like the conservation of matter or the laws of thermodynamics. It was just an empirical observation. But "double every year" is a memorably simple rule with incredible implications.

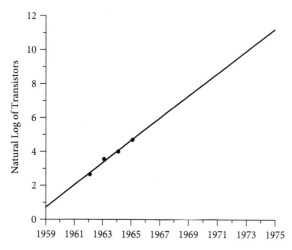

Figure 3.6 Doubling every year.

Moore noted that a doubling every year would increase the number of transistors per square inch between 1965 and 1975 by an astonishing factor of 1000, to 65,000, and claimed that it was, in fact, possible to cram this many transistors onto a circuit:

Certainly over the short-term this rate can be expected to continue, if not to increase. Over the longer term, the rate of increase is a bit more uncertain, although there is no reason to believe it will not remain nearly constant for at least 10 years. That means by 1975, the number of components per integrated circuit for minimum cost will be 65,000.

I believe that such a large circuit can be built on a single wafer.

The actual number in 1975 turned out to be about one-tenth Moore's prediction, leading Moore to revise his calculation of the growth rate from a doubling every year to a doubling every two years.

Figure 3.7 shows that the number of transistors per inch has continued to increase astonishingly for more than fifty years, on average, doubling every two years. Moore's provocative, but incorrect, prediction that it would become possible to cram 65,000 transistors onto a circuit in 1975 has been dwarfed by nine billion transistors per square inch in 2018.

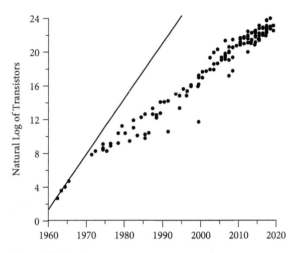

Figure 3.7 Doubling every two years.

In a 2015 interview, Moore said that, "I guess I see Moore's law dying here in the next decade or so, but that's not surprising." In 2016, a top Intel engineer said that Moore's law might end in four or five years. Many see future chips as being much smarter, not much smaller.

Moore's law is a simple, overly precise, statement of a general pattern. It *is* amazing how, year after year, computer components have become much more powerful and less expensive, but it is a mistake to think that this progress can be described by a simple mathematical law.

Patterns without reasons are unreliable.

How to Avoid Being Misled by Phantom Patterns

We are hard-wired to notice, seek, and be influenced by patterns. Sometimes these turn out to be useful; other times, they dupe and deceive us. Our affinity for patterns is so strong that it survived the Age of Enlightenment and the victory of the scientific method—no doubt, aided and abetted by selective recall and confirmation bias. We remember when a pattern persists and confirms our belief in it. We forget or explain away times when it doesn't.

We are still under the spell of silly superstitions and captivated by numerical coincidences. We still think that some numbers are lucky, and others unlucky, even though the numbers deemed lucky and unlucky vary from culture to culture. We still think some numbers are special and notice them all around us. We still turn numerical patterns into laws and extrapolate flukes into confident predictions.

The allure of patterns is hard to ignore. The temptation is hard to resist. The first step is to recognize the seduction.

CHAPTER 4

Fooled Again and Again

We have inherited from our distant ancestors a mistaken view of randomness. We think that random events, like coin flips, should alternate and not have streaks of several heads or tails in a row. So, a coin flip that lands heads must soon be followed by a tails. More generally, life's ups must be followed by downs, highs by lows, good by bad. If that were true, these events would not be random since heads would make tails more likely—which is not random at all. When something is truly random, streaks can, and do, happen.

Flippers and Fakers

Gary and Jerry were asked to flip a coin ten times and record the results, which are shown in Table 4.1. One person followed the rules and flipped a coin ten times. The other didn't bother with a coin and wrote down ten imaginary coin flips. Who was the flipper and who was the faker?

Another way to visualize these data is shown in Figure 4.1. Which sequence of flips looks more random to you?

Most people who look at these results think that Jerry is the flipper and Gary is the faker. Jerry reported fifty percent heads, while Gary reported seventy percent. Gary reported a streak of four heads in a row and another streak of three heads in a row, while Jerry's longest streaks were two heads in a row and two tails in a row.

Despite appearances, Gary's flips are real. We know this because he flipped a coin ten times in a recent statistics class and the results are shown in Table 4.1. We know that Jerry's flips are fake because we told him to imagine ten coin flips.

Table 4.1 *Ten flips and fakes.*

Gary	Jerry
H	H
H	T
H	T
T	H
T	T
H	H
H	H
H	T
H	H
T	T

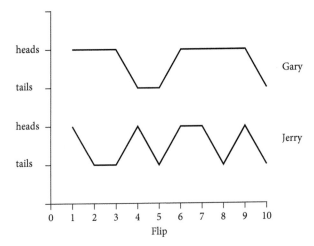

Figure 4.1 Can you spot the fake?

Gary's real flips are not unusual. Jerry's imagined flips are quite unrealistic. There is only a twenty-five percent chance that ten flips will yield five heads and five tails. Three out of four times, there will be an imbalance one way or another.

As for streaks, Gary's real results are much more likely than Jerry's fake results. In ten coin flips, there is only a seventeen percent chance that the longest streak of consecutive heads or tails will be two, while there is a forty-six percent chance of a streak of four or more. Gary's actual longest streak of four is much more likely than Jerry's reported longest streak of two.

The more data we look at, the more likely we are to find streaks and other striking patterns. If we flip a coin ten times, it is very unlikely that we will get a streak of ten heads or ten tails. But if we flip a coin 1,000 times, there is a sixty-two percent chance that there will be a streak of ten or more consecutive heads or tails somewhere in those flips. Patterns that seem unlikely are actually very likely—especially with lots of data.

The misperception that random data don't have streaks can cause all sorts of mischief. When we see a streak, it is tempting to leap to the conclusion that that something real is going on. If a gambler wins four games in a row, we think that he must be hot and is very likely to keep winning. If a stock picker makes four correct predictions in a row, we think that she must be a guru worth paying for advice. In reality, they both were probably lucky. The more games we watch and the more stock pickers we follow, the more likely it is that someone will have a lucky streak.

Spotify, iTunes, and other digital music players offer a shuffle mode in which a built-in algorithm randomly selects songs from the user's playlist. When they were first introduced, the algorithms were truly random in that every song on the playlist had an equal chance of being selected to be played next. With so many people playing so much music, there were bound to be streaks in which the same artist or genre was played several times in a row—leading to complaints that something was wrong with the algorithm.

There was nothing was wrong with the algorithms, only with users' perceptions of what random selections look like. The companies decided to modify their algorithms so that, instead of being truly random, the order in which songs are played looks more like what users expect.

Drunken Steps

Table 4.1 and Figure 4.1 show Gary's coin flips—heads or tails. Figure 4.2 shows the cumulative difference between the number of heads and tails. It doesn't look random at all, even though these were ten perfectly ordinary coin flips. It seems that the difference between the number of heads and tails is on an upward trend, portending a growing imbalance between heads and tails. But remember, these are coin flips, and that's what happens

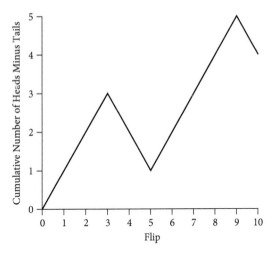

Figure 4.2 Cumulative difference between the number of heads and tails for Gary.

with random events like coin flips. The outcomes can, by chance, run in one direction or another for extended periods of time.

Figure 4.2 depicts what is known as a *random walk*. At any point in time, the difference between the number of heads and tails may go up (if the coin lands heads) or down (if the coin lands tails). The walk is random because the direction it moves is independent of previous movements, just like the next step made by a drunkard could be in any direction and is independent of previous steps.

The paradoxical thing about random walks is that, although each step is random and independent of previous steps, the walk can, by luck alone, wander off in one direction or another, and form striking patterns. For example, many technical analysts study charts of the prices of precious metals, stocks, and other investments, hoping to predict which way prices will go next. Even if the prices follow a random walk—and are therefore unpredictable—patterns are inevitable.

An "open-high-low-close" price chart for the daily prices of gold and other investments uses a sequence of daily vertical lines, with each line spanning the low and high prices that day, a hash mark on the left side of the vertical line showing the opening price that day, and a hash mark on the right side showing the closing price.

Figure 4.3 and Figure 4.4 show two open-high-low-close charts for daily gold prices over a 100-day period. One of these charts is a real graph of gold prices; the other is a fake chart constructed by starting at a price of $1000 and then, every imaginary day, flipping an electronic coin twenty-five times, with the price going up when the coin landed heads and down when the coin landed tails. This experiment was repeated 100 times, corresponding to 100 trading days for these imaginary gold prices.

Which chart, Figure 4.3 or Figure 4.4, is real and which is fake? We didn't put any numbers on the vertical axis, because we didn't want to give any clues. The question is whether someone looking at these two open-high-low-close charts can confidently distinguish between real and fake gold prices.

The answer is no, and that is the point. When technical analysts study charts like these, they often find patterns (like upward channels and support levels) that they think are meaningful. What they don't appreciate fully is that even random walks, in which future price movements are completely independent of past price changes, can generate patterns. If we can't tell whether a pattern came from real prices or from random coin flips, then it cannot possibly be useful for making price predictions.

(BTW: Figure 4.3 is real; Figure 4.4 is fake.)

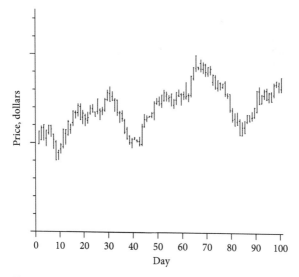

Figure 4.3 Gold prices (or coin flips) trading in an upward channel.

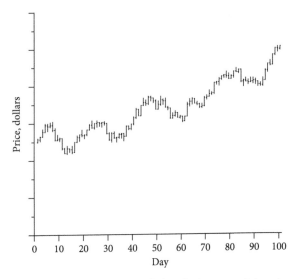

Figure 4.4 Gold prices (or coin flips) trading in an upward channel.

Warm Hands

Many athletes and fans believe that players sometimes get *hot hands*, with the chances of success temporarily elevated, or *cold hands*, with the chances of success temporarily deflated. For example, they see a basketball player who normally makes fifty percent of his shots get hot and make several shots in a row. Purvis Short, a National Basketball Association (NBA) player who once scored fifty-nine points in an NBA game, argued that, "You're in a world all your own. It's hard to describe. But the basket seems to be so wide. No matter what you do, you know the ball is going to go in."

Are hot and cold hands real, or an illusion? We know that there is a forty-six percent chance that a coin flipped ten times will have a streak of at least four consecutive heads or four consecutive tails, and we know that such streaks are meaningless because the chances of heads or tails on the next flip remain a rock-solid fifty-fifty. Are athletic performances the same way—temporary streaks that are nothing more than meaningless coincidence?

This question is difficult to answer because, unlike coin flips, conditions change constantly during most athletic competitions. In an NBA game, a player might attempt a one-foot layup on one play and a twenty-four-foot jump shot on the next play, or be guarded tightly on one play and loosely on the next.

Gary has done studies of bowling and horseshoes, which have stable conditions, and concluded that, although players may not get hot hands, they do get warm hands in that the chances of rolling strikes and pitching ringers increase modestly after previous successes. This may be because, unlike coins, humans have memories and often perform better when they are confident.

It is plausible that self-confident athletes perform better than doubters, and there is experimental evidence to support this idea. One study measured the arm strengths of twenty-four college students who were then paired up in arm-wrestling competitions with both opponents given incorrect information about who had greater arm strength. The weaker person won ten of twelve matches. In a similar experiment involving a muscular leg-endurance competition, the student participants performed better when they were told (incorrectly) that their opponent had recently had knee surgery than when they were told (incorrectly) that their opponent was a varsity track athlete.

Gary looked at bowling and horseshoes because the rolls and pitches are made under similar conditions with relatively little time between attempts. There are also relatively stable conditions at the NBA's annual three-point shooting contest. The NBA All-Star game is a mid-season exhibition game involving two dozen of the league's best players. The emphasis is on spectacular offensive plays that entertain the fans, with the defense mainly trying to stay out of the way. In 2019, the final score in the All-Star Game was 178–164, compared to an average of 111 points per game during the regular season.

In addition to the All-Star Game, several other events happen during All-Star Weekend, including a slam dunk contest and a three-point shooting contest. The three-point contest involves invited players taking turns attempting twenty-five long-range shots, separated into five shots from each of five stations. The top three shooters from a qualifying round move on to the championship round, where they again attempt twenty-five shots. (Before 2000, the contest had three rounds.)

There were ten participants in the 2019 contest, with Steph Curry, Joe Harris, and Buddy Hield advancing to the championship round. Harris won, making nineteen of twenty-five shots, including an astounding twelve in a row. The buzzword on social media was *hot*. Harris definitely got hot. Or did he? If a player makes several shots in a row, this isn't necessarily evidence that he is streaky. Maybe he is just a very good shooter.

Harris is a terrific three-point shooter. He led the league during the regular season, making 47.4 percent of his three-point shots. The All-Star three-point competition is easier than the regular season because there are no defenders. Harris made thirty-six of fifty (seventy-two percent) of his attempts in the qualifying and championship rounds. However, the probability that anyone, even someone as good as Harris, would make twelve out of twelve shots is very low. In Harris' case, assuming that every shot has a seventy-two percent success rate, the probability that he would make twelve of twelve shots is 0.019, or less than two percent.

The standard hurdle for statistical significance in scientific studies is five percent. If something happens that has less than a five percent chance of happening by chance alone, then we are justified in concluding that more than mere chance is involved. Harris's streak was evidently statistically significant evidence that he got hot.

Not so fast. In the championship round, Harris took twenty-five shots and his streak of twelve in a row started with his third shot. We should calculate the probability that he would make twelve in a row at some point in these twenty-five shots, not the probability that, if he took only twelve shots, he would make all twelve.

We can calculate the probability of Harris making twelve in a row at some point in a twenty-five-shot round by considering the fact that he made nineteen shots and missed six. If his twenty-five shots were independent, with no hot or cold periods, every possible arrangement of the nineteen made shots and six missed shots would be equally likely. Figure 4.5 shows the probabilities of streaks of various lengths. The chances of a streak of at least twelve is 6.8 percent, not quite statistically significant by the five-percent rule.

Even this 6.8 percent calculation is misleading. There were ten players and thirteen sets of twenty-five shots (ten qualifying sets and three championship sets). It is cherry-picking to focus on Harris' championship set after the results are in, and pretend that we didn't look at the other twelve sets.

Table 4.2 shows the comparable calculations for each player in the qualifying round and for the three players in the championship round.

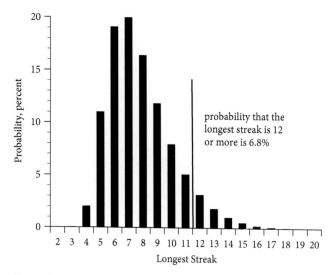

Figure 4.5 Longest-streak probabilities for Joe Harris' championship round.

Harris' streak is the most remarkable. For most players, the longest streaks are distinctly unremarkable.

To take into account that there was a total of thirteen sets, the correct question is the probability that, by chance alone, at least one player would have a streak with less than a five percent chance of occurring. That probability is a sobering forty-nine percent. If every shot were independent, with no hot or cold hands, there is a forty-nine percent chance that at least one player would have a streak that, by itself, has less than a five percent chance of occurring.

This is precisely why it is misleading to look for a pattern after the data have been collected. Here, it is misleading to look at all thirteen sets and pick out Harris' championship round as compelling evidence that basketball players have hot streaks.

When we consider the entirety of the data for the 2019 contest, it is as likely as not that at least one player would have a statistically impressive streak. The fact that the best we found was one streak that not quite statistically significant is, if anything, evidence against the claim that basketball players get hot hands.

Table 4.2 NBA All-Star three-point shooting contest, 2019.

Player	Team	Shots Made	Longest Streak	Probability
Qualifying Round				
Joe Harris	Brooklyn Nets	17	8	0.202
Kemba Walker	Charlotte Hornets	11	4	0.814
Khris Middleton	Milwaukee Bucks	8	7	0.359
Seth Curry	Portland Trail Blazers	13	4	0.766
Damian Lillard	Portland Trail Blazers	13	4	0.766
Buddy Hield	Sacramento Kings	18	7	0.511
Danny Green	Toronto Raptors	16	5	0.714
Dirk Nowitzki	Dallas Mavericks	12	5	0.412
Steph Curry	Golden State Warriors	19	10	0.198
Devin Booker	Phoenix Suns	15	7	0.149
Championship Round				
Joe Harris	Brooklyn Nets	19	12	0.067
Buddy Hield	Sacramento Kings	13	5	0.412
Steph Curry	Golden State Warriors	18	9	0.190

The calculations we have done so far consider whether a player gets hot *during* a round, in Harris' case, stringing together twelve of the nineteen shots that he made in the championship round. Another way a player might get hot is by doing better in one round than in his other rounds; for example, shooting fifty percent in five rounds over three years of contests, but hitting eighty percent in one fiery round. For Joe Harris, 2019 was his first contest and he shot seventeen of twenty-five in the first round with an eight-shot streak and nineteen of twenty-five in the championship round with a twelve-shot streak. He was, if anything, consistent, but we can't conclude much from two rounds of data.

What about other players in the history of the contest? The NBA three-point competition has been going on since 1986 and 124 players have taken a total of 389 twenty-five-shot sets. Arguably the greatest performance was in 1991 when Craig Hodges made twenty-one of twenty-five shots in the semi-final round, including an astonishing nineteen in a row—both all-time records. Hodges also shot very well in the first round of the very first three-point contest in 1986, when he made twenty of twenty-five shots.

Hodges played in the NBA for ten years and led the league in three-point accuracy three times. His career three-point average was forty percent. Hodges participated in the three-point contest eight times, winning three times and finishing second twice. He was a great three-point shooter and we have nineteen rounds of contest data for him.

We know the details of his 1991 semi-final round, but we don't have a complete record for several of his other rounds. Some of the twenty-five balls used in the contest are red-white-and-blue "money balls" that are worth two points, compared to one point for the standard balls. For many of the early contests, the recorded results show each player's point score, but not the number of shots made or the sequence in which they were made.

A three-point fanatic contacted Hodges directly, trying to unearth information about his 1986 performance, and got this response:

Peace…its Craig Hodges…not sure where u can find that video…likewise not sure how many shots I made in the very first round of the first contest ever…Not sure if that helps…Peace Hodge

We were able to find a newspaper story reporting that Hodges made twenty of twenty-five shots in the first round in 1986, but we could not find information about several other contests.

Overall, Hodges scored 321/570 = 0.563 possible points in his nineteen rounds. If we assume that he made 56.3 percent of his 475 total shots, that works out to be 267.5 shots made, which we rounded to 268. Our first set of questions, then, is for a player who made 268 of 475 shots in nineteen rounds, what is the probability that he would (a) have at least one round where he made twenty-one of twenty-five shots, and (b) have at least one round where he made nineteen shots in a row? Our second set of questions is, taking into account that 124 players have participated in the contest, what is the probability that at least one player would do something as unusual as Hodges did?

The answers are shown in Table 4.3. First, given Hodges' overall performance, there is a low probability that he would do as well as he did in his 1991 semi-final round, particularly his nineteen-shot streak. Second, given that 124 players have participated in the contest, there is a low probability that any player would do something as unlikely as Hodges' nineteen-shot streak.

Table 4.3 *Hodges got hot in his 1991 semi-final round.*

Result	Probability that Hodges Would Do This Well	Probability that Someone Would Do Something This Unusual
21 of 25	0.0026716	0.2823
19 in a row	0.0000477	0.0059

One important conclusion is that it is certainly not a straightforward task to assess whether basketball players have hot and cold streaks! Superficial evidence, such as a player making several shots in a row, may well be due to chance—in the same way that a coin might land heads several times in a row. Even when we identify something truly remarkable, like Joe Harris making nineteen of twenty-five shots, including twelve in a row, in 2019, it might simply be explained by the fact that we cherry-picked that performance—like flipping ten coins 1,000 times and noting that all ten coins once landed heads.

However, there are statistical ways to account for cherry-picking and, when we do so, Craig Hodges' streak of nineteen in a row in the 1991 contest is too improbable to be explained by luck or cherry-picking. He was hot.

Hot hands don't happen every day, but they do happen.

Are You Picking on Me?

In 2018, Eric Reid, a Carolina Panthers football player, complained that the National Football League (NFL) was using its purportedly random drug-testing program to target him. He had been tested seven times in eleven weeks, and an easy explanation was that the NFL was picking on him because he had been the first player to join Colin Kaepernick in kneeling during the national anthem and he had also been fined several times for excessively violent tackles.

Reid said that, "That has to be statistically impossible. I'm not a mathematician, but there's no way that's random." One commentator agreed: "The odds of any one player being 'randomly' tested [that many times] are incredibly low." Reid's coach supported his player: "If my name came up that many times, I'd buy a lottery ticket."

Soon other players chimed in, complaining that they had been tested after making a violent tackle or a dumb joke. Yahoo Sports reported that the probability that Reid would be selected so often was a mere 0.17 percent.

This situation is just like coin flips and three-point shots. Before the flips or shots begin, the probability that a specific pattern will occur is small but, after the flips or shots have happened, the probability that there will be some pattern is high. At the start of the season, the probability that Eric Reid would be tested multiple times is small but, after eleven weeks of testing, the probability that someone will have been tested multiple times is large.

An independent laboratory uses a computer algorithm to select the names of the players who will be tested, and the NFL and the NFL players union both investigated Reid's complaint and concluded that Reid's tests were indeed random.

Let's check the probabilities. Reid's first drug test was a mandatory test after he signed with the Panthers. After that initial test, ten players on each NFL team are randomly selected each week and Reid was chosen six times in eleven weeks. If we had picked out Reid before the season started, the probability that his name would come up six or more times is 0.001766. (This is the probability that Yahoo Sports reported.)

However, we didn't do that. We picked out Reid's name *after* eleven games had been played. There are seventy-two players eligible for testing on the Carolina roster, any one of whom could have been selected multiple times. The chances that *someone* will be selected at least six times is higher than the chances Reid will be chosen six times. Specifically, the probability that at least one Carolina player would be selected six or more times is 0.119. In addition, there are thirty-two NFL teams and the chances are pretty good that some player on some team will be selected six times. The probability that at least one NFL player would be tested at least six times works out to be 0.983—a near certainty.

As with many random events, it is easy, after the fact, to find patterns that would have been difficult to predict ahead of time. We are likely to find something afterward that was unlikely beforehand.

Slow Down and Shuffle

Bridge is a game played with a standard deck of fifty-two cards that are shuffled and dealt to four players who are on opposing two-player teams. During the play of the hand, each "trick" consists of a designated player leading a card and the other players following suit if they can. For example, if a player leads the five of spades, the other players must play a spade, too, if they have any.

There are so many possible hands that can be dealt that, in practice, every hand is different—which makes for an endlessly challenging and entertaining game that involves complex and subtle strategies. Indeed, bridge is one of the few games where computer algorithms have not yet defeated the best human players.

Bridge hands used to be shuffled and dealt by the players. In the late 1970s and early 1980s, serious competitions began switching to machine-shuffled hands. At first, players complained that the machines were faulty because they dealt too many wild hands with uneven distributions. More often than they remembered, at least one player was dealt a void (no cards in one suit) or six, seven or more cards in a suit.

These complaints were taken seriously because the people who played in these competitive matches had years and years of experience to back up their claims that the machine-shuffled hands had wilder distributions than the hands they were used to. Several mathematicians stepped up and calculated the theoretical probabilities and compared these to the actual distribution of machine-shuffled hands. It turned out that the distribution of machine-shuffled hands was correct. For example, eighteen percent of the time, at least one player should have a void; fifty percent of the time, at least one player should have six or more cards in the same suit; and a remarkable fifteen percent of the time, at least one player should have seven or more cards in the same suit. The machine-shuffled hands matched these frequencies.

The problem was not with the shuffling machines, but with human shufflers.

As with coin flips, bridge players did not appreciate how often randomly selected cards show seemingly unusual patterns, and this disbelief was reinforced by years and years of inadequate shuffling.

When a bridge hand is played, many tricks have four cards of the same suit; four spades, for example. In addition, the same suit is often led two or three times in a row, causing eight to twelve cards in the same suit to be bunched together. When the cards are collected at the end of a hand, many cards in the same suit are likely to be clustered together. If the deck is shuffled only two or three times in order to get on to the next hand ("Hurry up and deal!"), some of those bunched suits will survive largely intact and be dealt evenly to each player. An extreme case would be where a trick with four spades is not broken up by two or three shuffles, guaranteeing that when the cards are dealt, each player will get one of these four spades—which makes it impossible to have a void in spades and difficult for any player to have seven or more spades. The problem was that humans were not shuffling the cards enough!

Persi Diaconis, a statistician and former professional magician, has shown—both in theory and practice—that if a deck of cards is divided into two equal halves and the cards are shuffled perfectly, alternating one card from each half, the deck returns to its original order after eight perfect shuffles. It is our imperfect shuffles that cause the deck to depart from its original order, and it takes several flawed shuffles to mix the cards thoroughly. Diaconis and another statistician, Dave Bayer, showed that two, three, four, or even five imperfect human shuffles are not enough to randomize a deck of cards. Their rule of thumb is that seven shuffles are generally needed. Six shuffles are not enough and more than seven shuffles doesn't have much effect on the randomness of the deck.

If we want random outcomes when we play bridge, poker, and other card games, we should slow down and shuffle seven times.

Settlers

The Settlers of Catan is an incredible board game created by Klaus Teuber, a German game designer. It has been translated into dozens of languages and tens of millions of games have been sold. The basic four-player board consists of nineteen hexagons (hexes) representing resources: three brick, four lumber, four wool, four grain, three ore, and one desert. Players accumulate resources based on dice rolls, card draws, trading, and the location of their settlements and cities. Part of its seductive appeal is that the hexes can be laid out in an essentially unlimited number of ways, and player strategies depend on how the hexagons are arranged. The rules are simple, but the strategies are complex and elusive.

The official rules of Settlers of Catan recommend that the resource hexes be shuffled, randomly placed face down on the board, and then turned over. Figure 4.6 and Figure 4.7 show two hex arrangements. Which Settlers board looks more random to you?

This is a two-dimensional version of Gary and Jerry's ten coin tosses, using nineteen hexes instead of ten coins. We have a deep-rooted tendency

Figure 4.6 An unbalanced settlers board.

to think that clusters are unlikely to occur randomly—whether it be four heads in a row or three lumber hexes in a row. It is a misperception to think that heads and tails or Settlers hexes should alternate.

Players are often dismayed to find that a random arrangement results in a triplet, like the three lumber resources in Figure 4.6. Since the board doesn't appear to be random, the players rearrange the hexes until they find a layout like Figure 4.7 that seems random.

As with coin streaks, randomly placed hexes often have striking coincidental patterns. Predicting a specific pattern beforehand is difficult. Detecting some pattern after the fact is expected. There is a twenty-nine percent probability that a randomly constructed Settlers board will have at least one triplet, as in Figure 4.6. There is only a four percent chance of a board like Figure 4.7 in which no adjacent hexes have the same resource.

The game is more fun if players accept the fact that clusters should be expected, instead of limiting the possibilities by misguided notions of what randomness looks like.

Figure 4.7 A truly random settlers board?

Cancer Clusters

This clustering principle applies to things a lot more serious than board games. Suppose that the nineteen Settlers of Catan hex locations are nineteen small cities, each with 100 residents, and that each person has a ten percent chance of developing invasive cancer before the age of sixty, irrespective of where he or she lives.

We used a computer random number generator to determine whether each imaginary person develops cancer before the age of sixty. Figure 4.8 shows the outcomes of our computer simulation. Our results were not unusual. With 100 people in each city and a ten percent chance of developing cancer, we expect, on average, ten people in a city to develop cancer. In our simulation, the city average is an unremarkable 10.63, which, as expected, is close to ten but not exactly ten. Nor is it remarkable that one city had sixteen people develop cancer and another city had only five. With nineteen cities, the probability that at least one city will have sixteen or more cancer victims is fifty-five percent, and the probability that at least

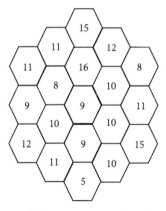

Figure 4.8 Cancer incidence in 19 small towns.

one city will have five or fewer is sixty-six percent. Like coin flips and Settlers of Catan, seemingly unusual outcomes are not unusual.

If these were real cities and we didn't appreciate how much variation occurs in random data, we might think that the cities with sixteen and five cancer cases are remarkable and we might try to find an explanation. We might also focus on the fact that two adjacent cities in the center north of the map had fifteen and sixteen cases, while the center south city had only five. With a little snooping, we might discover that there is a cell tower in the northern part of this region, and conclude that living near this cell tower causes cancer, and living far from the tower reduces the chances of developing cancer.

If there were more towns, even more extreme results would be likely because of nothing more than the fickle nature of luck. For example, with 1,000 towns, there is a ninety-two percent chance that at least one city will have twenty or more cancer victims, and an eighty percent chance that at least one city will have two or fewer. If we saw one city with twenty cases and another with only two cases, it would be tempting to search for an explanation—ways in which these cities differ—and we would surely find differences—perhaps in schools, parks, trees, water towers, or power lines—that seem important, but aren't.

In the 1970s there was, in fact, a much-ballyhooed report that exposure to electromagnetic fields (EMFs) from power lines cause cancer, based on an epidemiologist's discovery that some of the homes lived in by people who had died of cancer before the age of nineteen were near power lines. The reality is that scientists know a lot about EMFs and there is no plausible theory for how power line EMFs might cause cancer. The electromagnetic energy from power lines is far weaker than that from moonlight and the magnetic field is weaker than the earth's magnetic field. Despite subsequent studies and experiments refuting the claim, many people still think power lines cause cancer.

Once the toothpaste is out of the tube, it is hard to put it back in.

Smaller is Better (and Worse)

A related statistical principle is that when a large data set is broken up into small groups of data, we are likely to find seemingly unusual patterns among these small groups. This is why observed differences among small groups, as in Figure 4.8, are often meaningless.

If we take one million coin flips and divide them into, say, 100,000 groups of ten flips, some ten-flip groups are likely to be entirely heads while other groups will be all tails. It would be a mistake to think that these clusters of heads and tails are due to anything other than random chance—to think, for example, that a coin with heads on both sides was used for some of the flips and a coin with tails on both sides was used for other flips.

Such disparities are more likely if we compare small groups—say, groups of ten—than if we compare large groups—say, groups of 1,000—because there is more variability in the outcomes of ten flips than in the outcomes of 1,000 flips.

This principle can be applied to many, many situations where there is a substantial element of luck in the outcome. For example, even if there is nothing inherently good or bad about small towns, chance outcomes are more likely to be extreme—good and bad—in small towns than in large towns. This is true of academic performances, crime statistics, cancer rates, and much more. Identifying the best or worst towns may really just be identifying the smallest towns.

Standardized Tests

From 1998 through 2013, California's Standardized Testing and Reporting (STAR) program required all public-school students in Grades 2 to 11 to be tested each year using statewide standardized tests. All schools were given an Academic Performance Index (API) score for ranking the school statewide as well as in comparison to 100 schools with similar demographic characteristics. The API scores were released to the media, displayed on the Internet, and reported to parents in a School Accountability Report Card.

The API scores ranged from 200 to 1000, with an 800 target for every school. Any school with an API below 800 was given a one-year API growth target equal to five percent of the difference between its API and 800. Thus, a school with an API of 600 had an API growth target of 610. The target for a school with an API above 800 was to maintain its API.

A school's API score was determined by the percentage of students in each of five quintiles established by nationwide scores for the tests. A truly average school that has twenty percent of its students in each quintile would have an API of 655, well below the state's 800 target. A Lake Woebegone school, with scores all above average and evenly distributed between the 50th and 99th percentile would have an API of 890. (Garrison Keillor, host of the radio program *A Prairie Home Companion*, described the fictitious town of Lake Woebegone as a place where "all the children are above average." This impossibility has been termed the "Lake Woebegone Effect" by educational researchers to identify the flaw in claims that all schools should perform above average.)

We collected API data for the 394 unified school districts that encompass grades kindergarten through eleventh grade. The average unified school district had 7,646 students. Then we looked at the five school districts with the highest API scores. These top-performing school districts were all below-average in size and averaged 2,592 students, which is one-third the size of the average school district. Looking at these high-performing districts, we might think that students in small school districts do better than students in large school districts.

On the other hand, if we look at the five school districts with the lowest API scores, they, too, were all below-average in size, averaging only 138 students. Even if we discard school districts with fewer than 100 students, the average size of the five districts with the lowest API scores was only 440.

So, which is it? Are small school districts better or worse than large school districts? We can't tell by looking at the top and bottom performing districts. All that does is confirm our observation that there is more variability among small school districts, so they typically have the most extreme results—the highest scores and the lowest scores.

For an overall measure, we might look at the correlation between API scores and district sizes. It turns out that this correlation is 0.01, essentially zero. Focusing on the best or worse performing school districts is definitely misleading.

Violent Crimes

Do you think that St. Louis, Detroit, and Baltimore are crime-infested U.S. cities? According to official FBI statistics, the five U.S. cities with the most violent crimes per capita are five towns you probably never heard of: Industry, Vernon, Tavistock, Lakeside, and Teterboro.

The reason you haven't heard much about these crime-plagued cities is the same reason they have the highest crime rates—they are very small, with 2017 populations of 204, 113, 5, 8, and 69 respectively. When a city has five residents, one crime gives it a crime rate of 20,000 crimes per 100,000 residents, which is ten times the crime rate in St. Louis, Detroit, or Baltimore.

Table 4.4 shows the crime rates in eight cities—the five cities with the highest crime rates and the three cities with reputations for high crime rates. Should we conclude that small cities are more dangerous than large cities? If we did, we would again be overlooking the statistical fact that the extremes are typically found in small data sets, because there is more variability in small data sets.

Table 4.4 *Unsafe cities, 2017.*

City	Number of Violent Crimes	Population	Crime per 100,000 Citizens
Industry, California	73	204	35,784
Vernon, California	35	113	30,973
Tavistock, New Jersey	1	5	20,000
Lakeside, Colorado	1	8	12,500
Teterboro, New Jersey	5	69	7,246
St. Louis, Missouri	6,461	310,284	2,083
Detroit, Michigan	13,796	670,792	2,057
Baltimore, Maryland	12,430	613,217	2,027

Let's look at the five cities with the lowest crime rates. No surprise, they are all small cities. More than 1,000 cities had no violent crimes at all. Six of these crime-free cities had fewer than 100 residents and their average size was 1,820, which is one-twelfth the average size of U.S. cities.

Are small cities more dangerous or less dangerous than large cities? Probably neither. The correlation between city size and the violent-crime rate is a negligible 0.003. This is just another example of the principle that there is typically more variation in small data sets than in large data sets.

We're Number 1 (or Maybe Number 2)

Most athletic competitions have a final match, game, or series that determines the champion. Soccer, cricket, and rugby have World Cup Finals every four years. In the United States, football, basketball, and baseball have their annual Super Bowl, NBA Finals, and World Series. Afterward, the winning team is celebrated by its players and fans, while the losing team is reminded that second place is just the first place loser.

Does one game really tell us which team is the champion and which is the first place loser? Remember how much variation there is in small samples. Suppose that, one year, Germany and France are the two best soccer teams in the world and if they were to play each other 100 times, Germany would win sixty-seven times and lose thirty-three times. (Supporters of France and other teams, please don't send us your complaints; this is purely hypothetical.)

The point is that, unless the better team wins 100 percent of the time, the weaker team has a chance of winning. Here, Germany is clearly the better team, and would demonstrate their superiority if they played each other 100 times. However, if France and Germany were to play a single game in the World Cup Finals, France has a one-third chance of winning. If France did win, supporters would insist that France is the better team even though a single game is far too small a sample to establish which team is better.

Remember, too, that every team has to go through preliminary matches in order to reach the championship game. The best team might not even make it to the finals! Brazil has won five soccer World Cups, but their 1982 team, which might have been their best team ever, was knocked out of the World Cup quarterfinals by a score of 3–2 against Italy. Poland in 1974, France in 1986, and England in 1990 are just three teams on a long list of great teams that didn't make the finals, let alone win the championship.

What about sports like baseball, where the final two teams play up to seven games with the first team to win four games crowned champion? An up-to-seven games series certainly gives us more information, but is still far from definitive.

The 2019 baseball World Series pitted the Houston Astros against the Washington Nationals. The Nationals had won ninety-three of 162 games during the regular season (fifty-seven percent), while the Astros had won 107 games (sixty-six percent). The Astros lineup included the American League's most valuable player (MVP) Alex Bregman, rookie-of-the year Jordan Alvarez, and two of the three finalists for the Cy Young Award for best pitcher, Gerrit Cole and Justin Verlander (Cole was the winner).

The Astros had beaten the mighty New York Yankees (winner of 103 games during the regular season) by four games to one to get to the World Series and were heavy favorites over the Nationals. The betting odds gave the Astros a sixty-eight percent chance of winning the World Series, the most lopsided odds since 2007.

The day before the World Series began, Gary wrote:

Who will win the World Series? I don't know, but I do know that baseball is the quintessential game of luck. Line drives hit right at fielders, mis-hit balls dying in the infield. Fly balls barely caught and barely missed. Balls called strikes and strikes called balls. Even the best batters make twice as many outs as hits. Even the best teams lose more than a third of their games.

This season, the Houston Astros had the highest win percentage (66%) in baseball, yet they lost two out of six games to Baltimore, which only won a third of their games—not because Baltimore was the better team, but because Baltimore was the luckier team those two games.

The Astros are one of the 10 best teams this season (along with the Yankees, Tampa Bay, Minnesota, Cleveland, Oakland, Atlanta, Washington, St. Louis, and the Dodgers), but who would win a 7-game series between any two of these teams? Your guess is as good as mine—perhaps better—but it is still only a guess.

Gary reminded readers of the 1990 Oakland A's, with league MVP Rickey Henderson, Cy Young winner Bob Welch, and Cy Young runner-up Dave Stewart. Their reliever Dennis Eckersley had a 0.61 earned run average (ERA), with seventy-three strikeouts and five walks (one intentional) in seventy-three innings. They also had Mark McGwire, Jose Canseco, Carney Lansford, and a half dozen other stars.

Playing in the American League West, the toughest division at the time, they won 103 games during the regular season, the third year in a row that they led the league. On the eve of the World Series, award-winning sports journalist Thomas Boswell wrote in *The Washington Post*, "Let's make this short and sweet. The baseball season is over. Nobody's going to beat the Oakland A's."

The A's lost the World Series in four straight games to the Cincinnati Reds, who had only won ninety-one games during the regular season. Chicago writer Mike Royko said that it happened because the A's had three ex-Chicago Cubs players on their roster. No, it happened because pretty much anything can happen in a seven-game series between two good teams.

How about the 1969 New York Mets? They entered the league in 1962 and lost a record 120 games. Over their first six years, they averaged fifty-four wins and 108 losses, a 33.3 winning percentage. They improved to seventy-three wins and eighty-nine losses in 1968 and then got a flukey 100 wins in 1969. In the World Series, they faced the Baltimore Orioles, who had won 109 regular season games with All-Stars everywhere, including Frank Robinson, Brooks Robinson, and Boog Powell. They had pitchers Mike Cuellar (23–11 with a 2.38 ERA), Jim Palmer (16–4 with a 2.34 ERA), and Dave McNally (20–7 with a 3.22 ERA). Relievers Eddie Watt, Pete Richert, and Dick Hall had ERAs of 1.65, 2.20, and 1.92.

Behind Cueller, the Orioles breezed through the first game of the World Series just as expected, winning 4–1. Then the Orioles lost the next four games. For the next fourteen years, the Mets were a distinctly mediocre team, winning forty-six percent of their games; but for one amazing season, they were the Miracle Mets.

What are the chances that the better team, by all objective measures, will lose the World Series? Surprisingly high. In a game like baseball, where luck is so important, a brief seven-game series tells us very little.

A team's win probability varies from game to game along with the starting pitcher and other factors, but we can get a pretty good estimate of the overall probability that a team will win a seven-game series by simply assuming a constant win probability for each team. Suppose that the Astros have a sixty percent chance of winningly any single game against the Nationals. This probability is generous, since the Astros won sixty-six percent of their games during the regular season, playing against average teams. But we will use a sixty percent win probability against the Nationals to give the Astros the benefit of the doubt.

Even with this generous assessment of the Astros, it turns out that there is a thirty percent chance that the Nationals would be World Champions. If the Astros have a more plausible fifty-five percent chance of winning each game, the Nationals have a forty percent chance of popping the champagne.

We can also turn this question around, and ask how much our assessment of the Astros would be affected by how well the team does in the World Series. Internet companies use a formula called Bayes' rule to estimate and revise the probability that you will like a certain product or service based on information they collect about you. We can do the same thing here, and revise our 0.60 estimate of the Astro's win probability based on how well they do in the World Series.

The answer is not much at all. The Astro's regular-season record doesn't guarantee that they will win a seven-game series against the Nationals, and winning or losing the World Series doesn't tell us much about how good they are. Even if the Astros lose four straight games, our best estimate of the probability of them beating the Nationals in another game only drops slightly, from 0.600 to 0.594. If the Astros win four games straight, their probability increases slightly to 0.604. Anything in between has even less effect. No matter how the World Series turns out, it should barely affect our assessment of the Astros.

That is the nature of the beast. In a game like baseball, where so much chance is involved, a seven-game series is all about the luck. The best team often loses, and winning the World Series tells us very little about which team is really better.

Gary ended his article:

Enjoy the Series; I know I will. But notice the luck and remember the 1969 Mets and the 1990 A's.

How did it turn out? The first two games were played in Houston, and the Nationals won both, leading some fair-weather pundits to claim that they knew all along that Washington was the better team. The truth is that the outcome of two games doesn't tell us much. Then the series moved to Washington and the Astros won the next three games. Now the fair-weather pundits said that the Astros were back on track. The truth is that the outcome of three games doesn't tell us much.

The series went back to Houston, and the Nationals won both games and the World Series. It was the first seven-game series in the history of any major sport where every game was won by the visiting team. There were plenty of lucky moments—hard-hit balls that were just close enough to a fielder to be caught, or just far enough away to not be caught; umpires making good calls and umpires making bad calls; minor and major injuries.

If the Astros and Nationals played another seven-games series, who would win? One of the few things we can say with certainty is that it will not be the Yankees. Another certainty is that when two good teams play once, or even seven times, this isn't enough to tell us which team is better—i.e. which team would come out on top if they played each other 100 times.

Enjoy the championship matches, but don't be so hard on the first place loser.

How to Avoid Being Misled by Phantom Patterns

Patterns are inevitable. Streaks, clusters, and correlations are the norm, not the exception. In addition, when data are separated into small groups, we should not be surprised to discover substantial variation among the groups.

For example, in a large number of coin flips, there are likely to be coincidental clusters of heads and tails. If the flips are separated into small groups, there are likely to be large fortuitous differences among the groups.

In nationwide data on cancer, crime, test scores, or whatever, there are likely to be flukey clusters. If the data are separated into smaller geographic units like cities, there are likely to be striking differences among the cities, and the most extreme results are likely to be found in the smallest cities. In athletic competitions between reasonably well-matched teams, the outcome of a few games is almost meaningless.

Our challenge is to overcome our inherited inclination to think that all patterns are meaningful; for example, thinking that clustering in large data sets or differences among small data sets is something real that needs to be explained. Often, it is just meaningless happenstance.

CHAPTER 5

The Paradox of Big Data

Introduction

A computer repair business hired a well-respected data analytics firm to create an artificial intelligence (AI) program that would tell its technicians which parts they should bring with them on service calls. The analytics firm digitized an enormous database of telephone recordings of the customers' service requests and then created an algorithm that would find the words most closely correlated with each type of repair.

It was a lot of work converting sounds to text, because computers are still not perfect at recognizing what we are trying to say when we mumble, slur, mispronounce, and speak with accents, but it was hoped that the enormous amount of textual data that they had assembled would yield useful results.

How did it work out? In the succinct words of the data analytics firm: "We failed miserably."

One problem was that many of the words people use in phone calls contain very little useful information. For example, consider this snippet of conversation:

Company: Hello. ABC Electronics. How can I help you?

Customer: Hi! Um, this is Jerry Garcia. My son's computer's busted, you know.

Company: What's wrong with it?

Customer: Yeah, uh, I called you a while back, maybe two or three months, I don't know. No, you know, I think it was like July. Anyway, you saved us, but, hmm, this is different.

Company: Can you tell me what's wrong?

Customer: He can't find some stuff, you know, that he saved, um, that he needs for school. He knows it's like there somewhere, but he doesn't know where.

[. . .]
Company: Can you give us your address?
Customer: Sure, um, 127 West Green. We're easy to find—uhh, it's like a block from the Methodist Church on Apple.

A computer analysis of these words is a daunting task. Words can be ambiguous and have multiple meanings. The word "saved" shows up twice, but is only relevant to the computer problem in one of these two occasions. Ditto with the word "find." The words "green" and "apple" might be related to the problem, but aren't. Computer algorithms have a devil of a time telling the difference between relevant and irrelevant information because they literally do not know what words mean.

The larger difficulty is that almost all of the words in this conversation have nothing whatsoever to do with the problem the customer wants fixed. With thousands of words used in everyday conversation, a computer algorithm is likely to find many coincidental relationships. For example, "easy" might happened to have been used unusually often during phone calls about sticky keyboards, while "know" happened to have been used frequently during phone calls about a printer not working. Since neither word had anything to do with the customer's problem, these coincidental correlations are totally useless for identifying the reasons behind future customer phone calls.

The data analytics firm solved this problem by bringing in human expertise—the technicians who go out on call to fix the problems. The technicians were able to identify those keywords that are reliable indicators of the computer problems that need to be fixed and the parts needed to fix those problems.

An unguided search for patterns failed, while computer algorithms assisted by human expertise succeeded.

This example is hardly an anomaly. It has been estimated that eighty-five percent of all big data projects undertaken by businesses fail. One reason for this failure is an unfounded belief that all a business needs to do is let computers spot patterns.

Data Mining

The scientific method begins with a falsifiable theory, followed by the collection of data for a statistical test of the theory. *Data mining* goes in the other direction, analyzing data without being motivated by theories—indeed, viewing the use of expert knowledge as an unwelcome constraint that limits the possibilities for discovering new knowledge.

In a 2008 article titled, "The End of Theory: The data deluge makes the scientific method obsolete," the editor-in-chief of *Wired* magazine argued that,

Petabytes allow us to say: "Correlation is enough." We can stop looking for models. We can analyze the data without hypotheses about what it might show. We can throw the numbers into the biggest computing clusters the world has ever seen and let statistical algorithms find patterns where science cannot.

In 2013, two computer science professors described data mining as a quest "to reveal hidden patterns and secret correlations." What they mean by "hidden patterns and secret correlations" are relationships that would surprise experts. They believe that, in order to find things that are unfamiliar to experts, we need to let computers data mine without being constrained by the limited knowledge of experts.

An unrestrained quest for patterns often combines data mining with what Nobel laureate Ronald Coase called *data torturing*: manipulating, pruning, or rearranging data until some pattern appears. Some well-meaning researchers believe that if data mining can discover new relationships, then we shouldn't limit its discovery tricks. Separate the data into subsets that have patterns. Discard data that keep a pattern from being a pattern. Combine data if that creates a pattern. Discard outliers that break a pattern. Include outliers that create a pattern. Keep looking until something—anything—is found.

In the opening lines to a foreword for a book on using data mining for knowledge discovery, a computer science professor wrote, without evident irony:

"If you torture the data long enough, Nature will confess," said 1991 Nobel-winning economist Ronald Coase. The statement is still true. However, achieving this lofty goal is not easy. First, "long enough" may, in practice, be "too long" in many applications and thus unacceptable. Second, to get "confession" from large data sets one needs to use state-of-the-art "torturing" tools. Third, Nature is very stubborn—not yielding easily or unwilling to reveal its secrets at all.

Coase intended his comment not as a lofty goal to be achieved by using state-of-the-art data-torturing tools, but as a biting criticism of the practice of ransacking data in search of patterns.

For example, after slicing and dicing the data, a company might discover that most of its best female software engineers prefer mustard to ketchup on their hot dogs. Yet, when they discriminate against ketchup-loving female job applicants in favor of mustard lovers, the new hires are distinctly mediocre. The explanation for this flop is that the mustard preference of the top female software engineers was discovered by looking at hundreds of traits and hundreds of ways to separate and arrange the data. There were bound to be some coincidental correlations with software engineering prowess. When a correlation is coincidental, it vanishes when applied to new engineers. It is useless.

Our previous book, *The 9 Pitfalls of Data Science*, gives numerous examples of the unfortunate results that people in academia and industry have obtained using data-mined models with no underlying theory. One business executive repeatedly expressed his disdain for theory with the pithy comment, "Up is up." He believed that when a computer finds a pattern, there does not need to be a logical reason. Up is up. Unfortunately, it often turned out that "up was down" or "up was nothing" in that the discovered patterns that were supposed to increase the company's revenue actually reduced revenue or had no effect at all.

It is tempting to believe that the availability of vast amounts of data increases the likelihood that data mining will discover new, heretofore unknown, relationships. However, the reality is that coincidental patterns are inevitable in large data sets and, the larger the data set, the more likely it is that what we find is coincidental. The paradox of big data is that we think the data deluge will help us better understand the world and make better decisions, but, in fact, the more data we pillage for patterns, the more likely it is that what we find will be misleading and worthless.

Out-of-Sample Data

The perils of data mining are often exposed when a pattern that has been discovered by rummaging through data disappears when it is applied to fresh data. So, it would seem that an effective way of determining whether a statistical pattern is meaningful or meaningless is to divide the original data into two halves—*in-sample data* that can be used to discover models, and *out-of-sample* data that are held out so that they can be used to test those models that are discovered with the in-sample data. If a model uncovered with half the data works well with the other half, this is evidence that the model is useful. This procedure is sensible but, unfortunately, provides no guarantees.

Suppose that we are trying to figure out a way to predict the results of Liverpool football games in the English Premier League, and we divide the 2018 season into the first half (nineteen in-sample games) and the second half (nineteen out-of-sample games). If a data-mining algorithm looks at temperature data in hundreds of California cities on the day before Liverpool matches, it might discover that the difference between the high and low temperatures in Claremont, California, is a good predictor of the Liverpool score.

If this statistical pattern is purely coincidental (as it surely is), then testing the relationship on the out-of-sample data is likely to show that it is useless for predicting Liverpool scores.

If that happens, however, the data-mining algorithm can keep looking for other patterns (there are lots of cities in California, and other states, if needed) until it finds one that makes successful predictions with both the in-sample data and the out-of-sample data—and it is certain to succeed if a sufficiently large number of cities are considered. Just as spurious correlations can be discovered for the first nineteen games of the Premiere League season, so spurious correlations can be discovered for all thirty-eight games.

A pattern is generally considered statistically significant if there is less than a five percent chance that it would occur by luck alone. This means that if we are so misguided as to only compare groups of random numbers, five percent of the groups we compare will be statistically significant!

Five percent is one out of twenty, so we expect one out of every twenty correlations to pass the in-sample test, and one out of 400 to pass both the in-sample and out-of-sample tests. A determined head-in-the-sand researcher who analyzes 10,000 groups of unrelated data can expect to find twenty-five correlations that are statistically significant in-sample and out-of-sample. In the age of big data, there are a *lot* more than 10,000 data sets that can be analyzed and a lot more than twenty-five spurious correlations that will survive in-sample and out-of-sample tests.

Out-of-sample tests are surely valuable; however, data mining with out-of-sample data is still data mining and is still subject to the same pitfalls.

Crowding Out

There is a more subtle problem with wholesale data mining tempered by out-of-sample tests. Suppose that a data-mining algorithm is used to select predictor variables from a data set that includes a relatively small number of "true" variables causally related to the variable being predicted as well as a large number of "nuisance" variables that are independent of the variable being predicted. One problem, as we have seen, is that some nuisance variables are likely to be coincidentally successful both in-sample and out-of-sample, but then flop when the model goes live with new data.

A bigger problem is that a data-mining algorithm may select nuisance variables instead of the true variables that would be useful for making reliable predictions. Testing and retesting a data-mined model may eventually expose the nuisance variables as useless, but it can never bring back the true variables that were crowded out by the nuisance variables. The greater the number of nuisance variables initially considered, the more likely it is that some true variables will disappear without a trace.

Multiple Regression

To illustrate the perils of data mining, we ran some Monte Carlo computer simulations (named after the gambling mecca) that used a computer's random number generator to create hypothetical data. By doing a very large number of simulations, we were able to identify typical and untypical outcomes.

One great thing about Monte Carlo simulations is that, because we created the data, we know which variables are causally related and which are only coincidentally correlated—so that we can see how well statistical analysis can tell the difference.

Table 5.1 reports the results of simulations in which all of the candidate explanatory variables were nuisance variables. Every variable selected by the data-mining algorithm as being useful was actually useless, yet data mining consistently discovered a substantial number of variables that were highly correlated with the target variable. For example, with 100 candidate variables, the data-mining algorithm picked out, on average, 6.63 useless variables for making predictions.

Table 5.1 *Simulations with no true variables.*

Number of Candidate Variables	Average Number of Variables Selected	In-Sample Correlation	Out-of-Sample Correlation
5	1.11	0.244	0.000
10	1.27	0.258	0.000
50	3.05	0.385	0.000
100	6.63	0.549	0.000
500	97.79	1.000	0.000

Table 5.1 also shows that, as the number of candidate explanatory variables increases, so does the average number of nuisance variables selected. Regardless of how highly correlated these variables are with the target variable, they are completely useless for future predictions. The out-of-sample correlations necessarily average zero.

We did another set of simulations in which five variables (the *true* variables) were used to determine the value of the target variable. For example, we might have 100 candidate variables, of which five determine the target variable, and ninety-five are *nuisance* variables. If data mining worked, the five meaningful variables would always be in the set of variables selected by the data-mining algorithm and the nuisance variables would always be excluded.

Table 5.2 shows the results. The inclusion of five true variables did *not* eliminate the selection of nuisance variables; it simply increased the number of selected variables. The larger the number of candidate variables, the more nuisance variables are included and the worse the out-of-sample predictions. This is empirical evidence of the paradox of big data:

It would seem that having data for a large number of variables will help us find more reliable patterns; however, the more variables we ransack for patterns, the less likely it is that what we find will be useful.

Notice, too, that when fewer nuisance variables are considered, fewer nuisance variables are selected. Instead of unleashing a data-mining algorithm on hundreds or thousands or hundreds of thousands of unfiltered variables, it would be better to use human expertise to exclude as many nuisance variables as possible. This is a corollary of the paradox of big data:

Table 5.2 *Simulations with five true variables.*

Number of Candidate Variables	Average Number of Variables Selected	In-Sample Correlation	Out-of-Sample Correlation
5	4.50	0.657	0.606
10	4.74	0.663	0.600
50	6.99	0.714	0.543
100	10.71	0.780	0.478
500	97.84	1.000	0.266

The larger the number of possible explanatory variables, the more important is human expertise.

These simulations also document how a plethora of nuisance variables can crowd out true variables. With 100 candidate variables, for example, one or more true variables were crowded out fifty percent of the time, and two or more variables were crowded out sixteen percent of the time. There were even occasions when all five true variables were crowded out.

The bottom line is straightforward. Variables discovered through data mining can appear to be useful, even when they're irrelevant, and true variables can be overlooked and discarded even though they are useful. Both flaws undermine the promise of data mining.

Know Nothings

An insurance company had created a huge database with records of every telephone, fax, or email interaction with its customers. Whenever a customer contacted the company, an employee dutifully checked boxes and filled in blanks in order to create a permanent digital record of the contact. This database had, so far, been used only for billing and other clerical purposes, but now the company was hopeful that a data analytics firm could data mine the mountain of data that had been collected and determine reliable predictors of whether a person who contacted the firm would buy insurance.

The analytics firm did what it was told and created an algorithm that pillaged the data. Since computers do not know what words mean, the algorithm did not consider what any of the boxes or blanks meant. Its task—which it did very well—was simply to find patterns in the boxes and blanks that could be used to make reliable predictions of insurance purchases.

The algorithm was spectacularly successful, or should we say *suspiciously* successful? It was able to predict with 100 percent accuracy whether a person who contacted the firm bought insurance. If it hadn't been perfectly successful, it might have been accepted and used without a thought, but 100 percent perfection made even the analytics firm suspicious.

It took a considerable amount of detective work to solve the mystery of this success. It hinged on a single field in the database, which had not been well-labeled but had three possible answers: phone, fax, or e-mail. If any one of these answers was positive, the customer bought insurance.

The analytics firm had to reverse-engineer the database in order to figure out the sequence of events that led to this field. It turned out that when customers canceled their insurance, the firm used this field to record whether the cancellation request had been sent by phone, fax, or e-mail. In order to cancel insurance, a customer had to have purchased insurance, so there was a 100 percent match between this variable and having bought insurance. It was a 100 percent useless pattern.

Immigrant Mothers and Their Daughters

The United States is said to be a land of opportunity, where people can make of their lives what they want—where the son of Jamaican immigrants can become Secretary of State, and a poor child from Hope, Arkansas, or the son of a Kenyan scholarship student can become President.

However, many people believe that there is little or no economic mobility in the U.S. Michael Harrington once wrote that "the real explanation of why the poor are where they are is that they made the mistake of being born to the wrong parents." Destiny is determined at birth. A 2003 *Business Week* article entitled "Waking Up from the American Dream" and a 2004 essay by Nobel Laureate Paul Krugman essay titled "The Death of Horatio Alger" both argued that economic mobility in the U.S. is a myth. Poor families are trapped in poor neighborhoods with broken families, bad schools, and a culture of low aspirations. It is very difficult to escape this vicious cycle of poverty and misery.

Economic mobility is particularly important for immigrants, who often incur enormous financial and social costs because they hope to make better lives for themselves and their children. Are the children of U.S. immigrants better off than their immigrant parents?

George Jesus Borjas, a professor at Harvard's John F. Kennedy School of Government, has been called America's leading immigration economist by both *Business Week* and *The Wall Street Journal*. Borjas argues that, in the past, "Within a decade or two of immigrants' arrival their earnings would overtake the earnings of natives of comparable socioeconomic background…the children of immigrants were even more successful than their parents." However, Borjas concludes that this is no longer true: "Recent arrivals will probably earn 20 percent less than natives throughout much of their working lives." One of his central arguments is that ethnic neighborhoods have cultural and socio-economic characteristics that limit the intergenerational mobility of immigrants.

Studies of the intergeneration mobility of immigrants, including studies by Borjas, generally compare the average income of first-generation male immigrants in one U.S. Census survey with the average income of second-generation males in a later survey. Each Census survey contains an enormous amount of data, but a little good data would be better than lots of bad data. There are several problems with this comparison of first-generation males in one census with the average income of second-generation males in a later Census.

First, the neglect of females is unfortunate, since daughters and sons may lead quite different lives. In the movie *Gran Torino*, the daughter of Hmong immigrants says that "the women go to college and the men go to jail." If true, then measures of upward mobility based on the experience of sons may not apply to daughters.

Second, these studies look at individual income, but household income is arguably more relevant because one important aspect of mobility is the extent to which people marry people from different socio-economic backgrounds.

Third, poor families tend to have more children, and this makes generational averages misleading. Suppose that there are two immigrant males, one a software engineer earning $380,000 and the other a gardener earning $20,000 in 2000. The first immigrant has one son who earns $760,000 in 2020; the second immigrant has four sons, each earning $40,000. Each son earns 100 percent more than his father, but the *average* income of the second generation is sixteen percent less than the average income of the first generation.

Instead of comparing a first-generation average with a second-generation average, we should compare parents with their children. A reasonable proxy for household income is the neighborhoods that people live in. Homes in neighborhoods that are safe and clean with attractive amenities (like good public schools) are more desirable and consequently more expensive. Because it is costly to move from home to home, a household's neighborhood may be a good proxy for its economic status.

Gary was given special access to California birth records that can be used to link immigrant mothers with their daughters at comparable stages of their lives (similar age and number of children) and to identify their neighborhoods. Unfortunately, there are no comparable data for immigrant fathers and their sons.

Approximately eighty-five percent of the grown daughters of foreign-born mothers live in different ZIP codes than did their mothers, and most daughters who change ZIP codes move to more affluent ZIP codes. In comparison to the daughters of white women born in California, the daughters of immigrant women have equal geographic mobility and more economic mobility. The gap between the economic status of the daughters of foreign-born mothers and the daughters of California-born mothers is less than half the size of the gap between their mothers' economic status.

We shouldn't be seduced by the volume of data. *Relevant* data are better than *plentiful* data.

Getting Over It

Many interesting questions can only be answered with large amounts of data collected by observing people's behavior on the Internet. Ideally, the data would be collected from randomized controlled trials, as with A/B tests of different web page designs. When this is not possible (and usually it isn't), it is best to begin with a clear idea of what one is trying to accomplish and then find reliable data that are well-suited for that purpose. Sifting through an abundance of data looking for interesting patterns is likely to end badly.

With two students, Gary worked on a study that had a clear purpose and focused on data that would serve that purpose. One of the students spent much of his time in college playing online poker. It wasn't a total waste of time, because he made some money and learned some valuable critical thinking skills. He later earned a PhD in economics and now works for Mathematica Policy Research. The other student also earned a PhD in economics and now works for the Federal Reserve.

While they were students at Pomona, they wrote a joint senior thesis (with Gary initially the supervisor and later a co-author) using online poker data to answer an interesting research question. Daniel Kahneman and Amos Tversky had once argued that a "person who has not made peace with his losses is likely to accept gambles that would be unacceptable to him otherwise." The poker-playing student had a personal interest in seeing if poker players who suffer big losses are lured into making excessively risky wagers in an ill-advised attempt to win back what they had lost. He could use this insight to fine-tune his play against big losers. A study of this question would also be a test of Kahneman and Tversky's assertion.

In the popular poker game Texas Hold 'Em, with $25/$50 stakes, each hand begins with the player sitting directly to the left of the dealer putting a small blind of $25 into the pot, and the player two seats to the left of the dealer putting in a big blind of $50. Each player is then dealt two "hole cards" that only they are allowed to see. The players who have not already put money in the pot decide whether to play or fold. To play, the players must either "call" the big blind ($50) or raise the bet above $50, forcing the other players to match the highest bet or fold.

After this initial round of betting, three community cards ("the flop") are dealt, which are visible to everyone and combined with each player's two hole cards to make a five-card hand. There is another round of betting, followed by a fourth community card ("the turn"), more betting, a fifth community card ("the river"), and a final round of betting. The remaining player with the best five-card hand, which can be made from their two hole cards and the five community cards, wins the pot.

Gary and these students couldn't very well spend time hanging around serious poker games, peeking at the players' cards, and recording how they played them. For legal and financial reasons, they couldn't set up experiments with real players and real money. However, the poker-playing Pomona student was intimately familiar with an online source of data. Full Tilt Poker was an online poker room launched in June 2004 with the involvement of a team of poker professionals. The company and its website were regulated by the Kahnawake Gaming Commission in Canada's Mohawk Territory. Because it was outside U.S. jurisdiction, the website was able to avoid U.S. regulations and taxes (although all that changed in 2011, after this study was completed).

The Full Tilt Poker site was great for Gary and the two students in that the players were risking real money and many of the games have large blinds that weed out novices. In fact, Full Tilt Poker boasted the largest online contingent of professional poker players anywhere. The clincher was that a computer program called PokerTracker allowed Gary and the students to record every online game—every card dealt, every wager, and every outcome. They recorded data twenty-four hours a day from January–May 2008 for all tables with $25/$50 blinds, which are considered high-stakes tables and attract experienced poker players. They ended up with data on the card-by-card play of hundreds of thousands of hands by knowledgeable people wagering serious money. These online data were literally the best way to test the theory that players change their style of play after big losses.

The poker-playing student knew that there is a generally accepted measure of *looseness*: the percentage of hands in which a player makes a voluntarily wager in order to see the flop cards. This can include a call or a raise, but does not include blind bets since these are involuntary.

Tight players fold when their two hole cards are not strong; loose players stay in, hoping that a lucky flop will strengthen their hand. At six-player tables, people are typically considered to be very tight players if their looseness is below twenty percent, and to be extremely loose players if their looseness is above fifty percent. For their data set, the average looseness at full six-player tables was twenty-six percent.

In theory, experienced poker players have a style that they feel works for them, which is based on hundreds of thousands of hands they have played. Once they have settled on their optimal strategy, they should stick to it, no matter what the outcome of the last few hands. If they suffer a big loss, they should recognize that it was bad luck and stick to their strategy. The research question was whether players who suffer big losses become less cautious and play hands they normally fold by putting money into the pot in order to see the flop.

Gary and the students considered a hand where a player won or lost $1,000 to be a significant win or loss. After a big win or loss, they monitored the player's behavior during the next twelve hands—two cycles around a six-player table. Their data set included 346 players who met the various criteria with the median number of hands played equal to 1,738. Half of the players won or lost more than $200,000, ten percent won or lost more than $1 million. As we said, these are real players wagering serious money.

The data consisted of six-player tables and heads-up tables which are limited to two players. There were sometimes empty seats at a six-player table, and this affects player strategies. For example, the chances that a pair of 8s in the hole will yield the best hand declines as the number of players increases. Gary and the students consequently grouped the data according to the number of players at the table. They did not combine the data for heads-up tables with the data for six-player tables with only two players because people who choose to play heads-up poker may have different styles than players who choose a six-player table but occasionally have four empty seats.

Gary and the students found that players indeed typically changed their style of play after winning or losing a big pot—most notably, playing less cautiously after a big loss, evidently hoping for lucky cards that will erase their loss quickly.

Table 5.3 shows that it was consistently the case that more players are looser after a big loss than after a big win. For example, with six players at the table, 135 players were looser after a big loss than after a big win, while the reverse was true for only sixty-eight players. To test the robustness of this conclusion, Gary and the students also looked at $250 and $500 thresholds for a big win or loss and found that in every case, most players play looser after a large loss. They also found that larger losses were evidently more memorable in that the fraction of the players who played looser after a loss increased as the size of the loss increased.

Was this change in strategy profitable? If experienced players are using profitable strategies to begin with, changing strategies will be a mistake. That's exactly what they found. Those players who played looser after a big loss were less profitable than they normally were.

Table 5.3 *Players were looser after a big loss than after a big win.*

Players at Table	Number of Players	Average Looseness	Players Who Were Looser After Big Win	Players Who Were Looser After Big Loss
heads-up	228	51	74	154
2	40	46	17	23
3	33	35	11	22
4	75	29	21	54
5	150	26	53	97
6	203	26	68	135

Because these researchers had a well-defined objective, they were able to focus on relevant data and reach a sensible conclusion that may well be applicable far beyond poker. We should all try to avoid letting setbacks goad us into making reckless decisions. We shouldn't make risky investments because we lost money in the stock market. We shouldn't replace a subordinate after one bad decision. We shouldn't rush into perilous relationships after a breakup.

Predicting an Exchange Rate

Gary once played a prank on a research assistant ("Alex") who insisted that the data speak for themselves: "Who needs causation when you have correlation?" Alex was exceptionally self-assured. He told Gary that when he died, he wanted everyone who had ever worked with him on group projects to carry his casket and lower it into the ground, so that they could let him down one last time.

Nothing Gary said dissuaded Alex from his unshakable belief that correlation is all that is needed to make reliable predictions. Alex argued that we don't need to know why two things are related and we shouldn't waste time asking why.

Finally, Gary proposed a wager. He would give Alex ten years of daily data for a target variable (unknown to Alex, it was the exchange rate between the Turkish lira and U.S. dollar) and data for 100 other variables that might be used to predict the target variable. To ensure that Alex did not use expert knowledge to build his model, none of the variables were labeled. The data would be divided into five years of in-sample data and five years of out-of-sample data and Alex could use whatever data-mining algorithm he wanted to see if he could discover a model that worked well with both in-sample and out-of-sample.

The only stakes were pride.

Gary gave Alex ten years of real data on the lira/dollar exchange rate but the other variables were just random numbers created by Gary. Alex constructed literally millions of models (75,287,520, to be exact) based on the first five years of data, and then tested these models with the second five years of data. His best model for the in-sample data had a 0.89 correlation between the predicted and actual values of the target variable, but the correlation for the out-of-sample data was –0.71! There was a strong positive in-sample correlation between the predicted and actual values and a strong negativwe correlation out-of-sample.

Alex was temporarily stunned that a model that had done so well in-sample could do so poorly out-of-sample. However, he had lots of data, so he kept data mining and was able to find several models that did exceptionally well both in-sample and out-of-sample. Figure 5.1 shows the in-sample and out-of-sample correlations for all 3,016 models that had an in-sample correlation above 0.85. Of these, 248 models had correlations above 0.80 out of sample, and ninety-two had correlations above 0.85. Many had higher correlations out-of-sample than in-sample!

Alex decided to go with the model that had the highest out-of-sample correlation. This model used variables 2, 7, 41, 53, 56 to predict the target variable and had a 0.87 correlation in-sample and a 0.92 correlation out-of-sample.

He proudly showed this model to Gary, confident that he had won the bet. However, Gary had a trick up his sleeve. He actually had fifteen years of data, from 2003 through 2017. Gary had anticipated that Alex would data mine, test, and repeat until he found a model that did well in-sample and out-of-sample, so he gave Alex the 2003–2007 in-sample data and 2008–2012 out-of-sample data, and reserved the remaining five years of

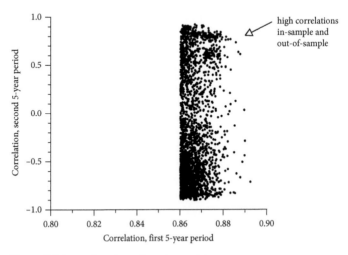

Figure 5.1 In-sample and out-of-sample correlations.

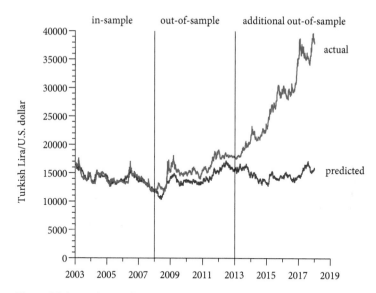

Figure 5.2 In-sample, out-of-sample, and additional out-of-sample.

data (2013–1017) for a fair test of Alex's model. Figure 5.2 shows that it flopped. The correlation between the predicted and actual value of the exchange rate was a paltry 0.08. Data do not speak for themselves.

Gary's second surprise was his revelation that all 100 of the candidate explanatory variables were random numbers, not economic data. Even though all of the in-sample correlations shown in Figure 5.1 were above 0.85 in the first five year period (2003–2007), the average correlation in the second five-year period (2008–2012) was 0.002, essentially zero. This makes sense because the predictor variables were literally random data and, no matter how well any of them did with the in-sample data, we expect them, on average, to be worthless out-of-sample.

By luck alone, 248 models happened to do well in both five-year periods. However, when these 248 models that had been constructed with in-sample data and validated with out-of-sample data were used for fresh data (2013–2017), the average correlation was again essentially zero.

Gary had two intended lessons. First, if enough models are considered, some are bound to appear to be useful, no matter how random (literally) the predictor variables. Second, predictive success in-sample and out-of-sample is no guarantee of success with new data.

Data Mining Trump's Tweets

In September 2019, there were 330 million active Twitter users and an average of 500 million tweets a day, nearly 200 billion a year. What a wonderful source of data to be ransacked and pillaged by data-mining algorithms!

Gary and John Zuk, one of Gary's former students, decided to focus on tweets sent by Donald Trump. He has sixty-six million Twitter followers and holds the most powerful office in the world, so perhaps his tweets have real consequences. Trump certainly thinks so. He has boasted of his "great and unmatched wisdom" and described himself as "great looking and smart, a true Stable Genius!" He shares his wisdom by sending an average of nine tweets a day.

Gary and John were not the first to data mine Trump's tweets. A Bank of America study found that the stock market does better on days when Trump tweets less. A JP Morgan study concluded that tweets containing the words *China*, *billion*, *products*, *Democrats*, and *great* have a statistically significant effect on interest rates.

Recognizing the importance of replication, Gary and John divided the three years following Trump's election victory on November 8, 2016, into the first two years (which were used for knowledge discovery) and the third year (which was used to see if the initial results held up out of sample). During this period, Trump sent approximately 10,000 tweets, containing approximately 15,000 unique words.

Daily fluctuations in Trump's word usage were correlated with the Dow Jones Industrial Average one to five days later. Many of these correlations did not hold up out of sample, but some did. For example, a one-standard-deviation increase in Trump's use of the word *thank* was predicted to increase the Dow Jones average four days later by 377 points. There was less than a one-in-a-thousand chance of a correlation as strong as the one that was discovered. Even better, the correlation during the out-of-sample period was even stronger than during the in-sample period!

Encouraged by this finding, Gary and John looked elsewhere. Trump has long admired Russian President Vladimir Putin. In June 2019, Trump said, "He is a great guy...He is a terrific person." Another time, he described their mutual admiration: "Putin said good things about me. He said, 'He's a leader and there's no question about it, he's a genius.'" Maybe Trump's tweets reverberate in Russia. Some data mining revealed that a one-standard-deviation increase in Trump's tweeting of the word *economy* was predicted to increase the high temperature in Moscow five days later by 2.00 degrees Fahrenheit. Again, the correlation was even stronger out-of-sample than in-sample.

Trump has also called North Korean Chairman Kim Jong-un a "great leader" and said that, "He speaks and his people sit up at attention. I want my people to do the same." This time, a little data mining discovered that a one-standard-deviation increase in Trump tweeting the word *great* was predicted to increase the high temperature in Pyongyang five days later by 2.79 degrees Fahrenheit. Once again, the correlation was even stronger out-of-sample.

Finally, Gary and John looked at the statistical relationship between Trump's choice of words and the number of runs scored by the Washington Nationals baseball team. CBS News reported that when Trump attended the fifth game of the 2019 World Series, there was "a torrent of boos and heckling from what sounded like a majority of the crowd" and chants of "Lock him up!" Perhaps the boos and heckling were because Nationals fans know that the fate of their team was determined by his tweets.

Sure enough, a data mining of Trump's tweets unearthed a statistically significant correlation between Trump's use of the word *year* and the number of runs scored by the Nationals four days later. A one-standard-deviation increase in the number of times *year* was tweeted increased the number of runs scored by the Nationals by 0.725.

Talk about knowledge discovery. We never anticipated the discovery of these statistically persuasive, heretofore unknown, relationships.

How to Avoid Being Misled by Phantom Patterns

The scientific method tests theories with data. Data mining dispenses with theories, and rummages through data for patterns, often aided by torturing the data with rearrangements, manipulations, and omissions. It is tempting to believe that big data increases the power of data mining. However, the paradox of big data is that the more data we ransack, the more likely it is that the patterns we find will be misleading and worthless.

Data do not speak for themselves, and up is not always up.

CHAPTER 6

Fruitless Searches

One of Gary's sons sometimes plays a game with his friends when they go to a restaurant for lunch. They all turn off their phones and put them in the center of the table. The first person to check his or her phone has to pay for lunch for everyone. The game seldom lasts long.

The loser of this game is never checking the weather, news, or stock prices. It is social media that has an irresistible lure. People want to see what other people are doing and share what they are doing. Checking our phones has been likened to a gambling addiction in that we are usually disappointed but, every once in a while, there is a big payoff that keeps us coming back, hoping for another thrill.

For data scientists, one of the payoffs from the Internet is the collection of vast amounts of social media data that they can scrutinize for patterns from the comfort of their offices and cubicles. No longer must they run experiments, observe behavior, or conduct surveys.

We are both data scientists and we appreciate the convenience of having data at our fingertips, but we are also skeptical of much of what passes for data these days. There is a crucial difference between data collected from randomized controlled trials (RCTs) and data consisting of the conversations people have on social media and the web pages they visit on the Internet.

An even more important problem, though, is that the explosion of data has vastly increased the number of coincidental patterns that can be discovered by tenacious researchers. If there are a relatively fixed number of useful patterns and the number of coincidental patterns grows exponentially, then the ratio of useful patterns to useless patterns must necessarily get closer to zero every day.

We illustrate our argument by looking at some real and concocted examples of using Google search data to discover useless patterns.

Google Trending the Stock Market

Burton Crane, a long-time financial writer for *The New York Times*, offered this seductive advice on how to make big money in the stock market:

Since we know stocks are going to fall as well as rise, we might as well get a little traffic out of them. A man who buys a stock at 10 and sells it at 20 makes 100 per cent. But a man who buys it at 10, sells it at 14 1/2, buys it back at 12 and sells it at 18, buys it back at 15 and sells it at 20, makes 188 per cent.

Yes, it would be immensely profitable to jump in and out of the stock market, nimbly buying before prices go up, and selling before prices go down. Alas, study after study has shown how difficult it is to time the market, compared to a simple buy-and-hold strategy of buying stocks and never selling unless money is needed for a house, retirement, or whatever else investors are saving money to buy.

Since stocks are profitable, on average, investors who switch in and out of stocks and are right only half the time will get a lower return than if they had stayed in stocks the whole time. It has been estimated that investors who jump in and out, trying to guess which way the market will go next, must be right three out of four times to do as well as they would following a buy-and-hold strategy. A study of professional investors by Merrill Lynch concluded that, relative to buy-and-hold, "the great majority of funds lose money as a result of their timing efforts."

These dismal results haven't stopped people from trying. In 2013, three researchers reported that they had found a novel way to time the market by using Google search data to predict whether the Dow Jones Industrial Average (Dow) was headed up or down. They used Google Trends, an online data source provided by Google, to collect weekly data on the frequency with which users searched for ninety-eight different keywords:

We included terms related to the concept of stock markets, with some terms suggested by the Google Sets service, a tool which identifies semantically related keywords. The set of terms used was therefore not arbitrarily chosen, as we intentionally introduced some financial bias.

The use of the pejorative word *bias* is unfortunate since it suggests that there is something wrong with using search terms related to the stock market. The conviction that correlation supersedes causation assumes that the way to discover new insights is to look for patterns unencumbered by what we think we know about the world—to discover ways to beat the stock market by looking for "unbiased" words that have nothing to do with stocks. The fatal flaw in a blind strategy is that coincidental patterns will almost certainly be found, and data alone cannot distinguish between meaningful and meaningless patterns. If we have wisdom about something, it is generally better to use it—in this case, to introduce some financial expertise.

For each of these ninety-eight words, the researchers calculated a weekly "momentum indicator" by comparing that week's search activity to the average search activity during the preceding weeks. This indicator is a *moving average* because it changes weekly as the range of weeks moves along. Table 6.1 gives an example, where the moving average is calculated over the preceding three weeks. (The search values provided by Google Trends are on a scale of 0 to 100 that reflects relative usage during the period covered.)

In week four of the stylized example in Table 6.1, the moving average over weeks one to three is 43, and the momentum indicator is $44 - 43 = 1$, which means that search activity for this keyword was above its recent three-week average. In week five, the three-week moving average is 44, and the momentum indicator is negative, $43 - 44 = -1$, because search activity in week five was below its recent average.

The researchers considered moving averages of one to six weeks for each of their ninety-eight keywords and they reported that the most successful stock trading strategy was based on the keyword *debt*, using a three-week moving average and this decision rule:

Table 6.1 *A momentum indicator using a three-week average.*

Week	Search Value	Three-Week Moving Average	Momentum Indicator
1	41		
2	43		
3	45		
4	44	43	1
5	43	44	−1

- Buy the Dow if the momentum indicator is negative;
- Sell the Dow if the momentum indicator is positive.

Using data for the seven-year period January 1, 2004, through February 22, 2011, they reported that this strategy had an astounding 23.0 percent annual return, compared to 2.2 percent for a buy-and hold strategy. Their conclusion:

Our results suggest that these warning signs in search volume data could have been exploited in the construction of profitable trading strategies.

They offer no reasons:

Future work will be needed to provide a thorough explanation of the underlying psychological mechanisms which lead people to search for terms like debt before selling stocks at a lower price.

The fatal problem with this study is that the researchers considered ninety-eight different keywords and six different moving averages (a total of 588 strategies). If they considered two trading rules (buying when the momentum indicator was positive *or* selling when the momentum indicator was positive), then 1,176 strategies were explored. With so many possibilities, some chance patterns would surely be discovered—which undermines the credibility of those that were reported.

Even the consideration of a large number of random numbers will yield some coincidentally profitable trading rules when we look backward—when we predict the past, instead of predicting the future.

Let's see how well their most successful model did predicting the future over the next seven years, from February 22, 2011 through December 31, 2018. Figure 6.1 shows the results. Their *debt* strategy had an annual return of 2.81 percent, compared to 8.60 percent for buy-and-hold.

It should be no surprise that their data-mined strategy flopped with fresh data. Cherry-picked patterns usually vanish.

We are not being facetious when we say that we sincerely hope that these researchers did not buy or sell stocks based on their data-mined strategy. On the other hand, a cynic reminded us that there's something about losing money that really forces people to take a hard look at their decision making.

Figure 6.1 Clumsily staggering in and out of the market.

Looking to the Stars

Instead of relying on Google keywords, why not tap into the wisdom of the ancients by predicting the Dow based on Google searches for the twelve astrological signs? Why not?

Jay is a Pisces. He says that explains why he doesn't believe in astrology. He doesn't think that stock prices are affected by the sun, moon, and planets—let alone Google searches for astrological signs. Nonetheless, we downloaded monthly Google search activity for the ten-year period 2009 through 2018 for the twelve astrological signs: Aries, Taurus, Gemini, Cancer, Leo, Virgo, Libra, Scorpio, Sagittarius, Capricorn, Aquarius, and Pisces. We then divided the data into a five-year in-sample period, 2009–2013, for creating the model, and a five-year out-of-sample period 2014–2018 for testing the model.

Figure 6.2 shows that a model based solely on searches for one astrological sign, Scorpio, worked great. The correlation between the predicted and actual values of the Dow is a remarkable 0.83.

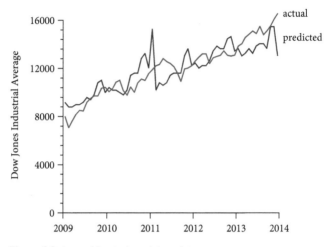

Figure 6.2 A promising stock market predictor.

Guess what? Gary is a Scorpio and he has invested and taught investing for decades!

If you thought, even for the briefest moment, that maybe this is why Scorpio was correlated with the Dow, you can understand how easily people are seduced by coincidental patterns. Gary does not cause stock prices to go up or down and Gary does not cause people to use Google to search for Scorpio more or less often.

One of the twelve correlations between stock prices and any twelve search words has to be higher than the other correlations and, whichever one it turns out to be, we can surely think of an explanation that is creative, but pointless.

Before rushing off to buy or sell stocks based on Scorpio searches, let's see how well the model did out-of-sample. Figure 6.3 shows that the answer, in one word, is *disappointing*. The stock market took off while search activity languished. Yet, again, we were fooled by a phantom pattern. It is easy to predict the past, even with nonsense models, but reliable predictions of the future require models that make sense.

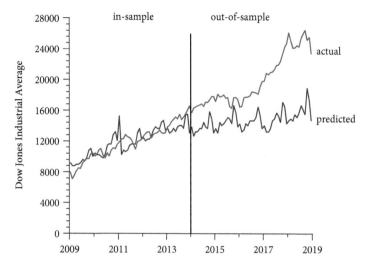

Figure 6.3 A disappointing stock market predictor.

Google Trending Bitcoin

Bitcoin is the most well-known cryptocurrency, a digital medium of exchange that operates independently of the central banking system. As an investment, Bitcoins are pure speculation. Investors who buy bonds receive interest. Investors who buy stocks receive dividends. Investors who buy apartment buildings receive rent. The only way people who invest in Bitcoins can make a profit is if they sell their bitcoins for more than they paid for them.

In 2018, Aleh Tsyvinski (a Yale professor of economics) and Yukun Liu (then a graduate student, now a professor himself) made headlines with a paper they wrote recommending that investors should hold at least one percent of their portfolio in Bitcoins. They also reported that Bitcoin prices could be predicted from how often the word *Bitcoin* is mentioned in Google searches. An increased number of *Bitcoin* searches typically precedes an increase in Bitcoin prices. A decline in *Bitcoin* searches predicts a decline in prices. They attributed this correlation to changes in "investor attention."

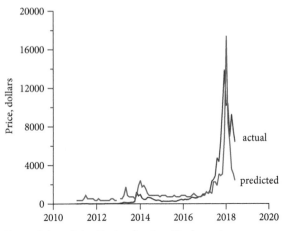

Figure 6.4 Predicting Bitcoin prices from Bitcoin searches.

For the time period they studied, January 2011 through May 2018, the correlation between monthly *Bitcoin* searches and the market price of Bitcoin on the first day of the next month was a remarkable 0.78. Figure 6.4 shows this close correlation, right up until the spring of 2018, when Bitcoin prices did not fall as much as predicted by the decline in *Bitcoin* searches.

If the documented correlation continued, this would be a novel way to get rich. In months when *Bitcoin* searches are high, buy on the last day of the month; in months when *Bitcoin* searches are low, sell on the last day of the month.

Part of the intellectual appeal of predicting Bitcoin prices from search data is that digital data are used in a novel way to predict the price of a digital currency.

The correlation also has a superficial appeal. Bitcoin prices are determined solely by what people are willing pay for Bitcoins and an increase in investor attention may well lead to an increased willingness to pay higher prices. On the other hand, many (if not most) people searching for information about this cryptocurrency may have no interest in buying. They are simply curious about this new thing called *Bitcoin*. In real estate, when a home is for sale, people who go to an open house, but have no intention of buying, are called lookie-loos. There are surely millions of Bitcoin lookie-loos, which makes Google searches a poor proxy for investor attention.

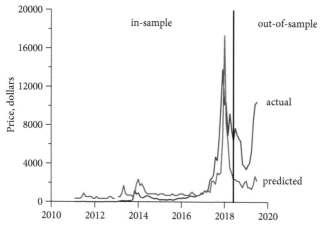

Figure 6.5 Mis-predicting Bitcoin prices from Bitcoin searches.

After the lookie-loos stop searching for more information about Bitcoins, will Bitcoin prices fall? Not if they weren't potential investors. Did the rise and subsequent fall in *Bitcoin* search activity shown in Figure 6.4 reflect investors chasing and then fleeing Bitcoin, or lookie-loos learning what they wanted to know about Bitcoin and then feeling they didn't need to learn much more?

Figure 6.5 shows what happened subsequent to the Tsyvinski/Liu analysis. Search activity stabilized while Bitcoin prices rebounded, causing search activity to be a terrible predictor of Bitcoin prices.

The original correlation had scant basis in theory. There was no compelling reason why there should be a lasting correlation between *Bitcoin* searches and Bitcoin prices. There was a temporary correlation, perhaps while lookie-loos searched for information about the speculative bubble in Bitcoin prices, but this correlation did not last and was useless for predicting future Bitcoin prices.

Figure 6.6 shows a comparable lookie-loo relationship between Bitcoin prices and searches for *Jumanji*, which refers to the movie *Jumanji: Welcome to the Jungle* that was released in December 2017, near the peak of the Bitcoin bubble. For the period January 2011 through May 2018, the correlation between *Jumanji* searches each month and the market price of Bitcoin on the first day of the next month is 0.73, which is comparable to the 0.78 correlation between *Bitcoin* searches and Bitcoin prices. Those who believe

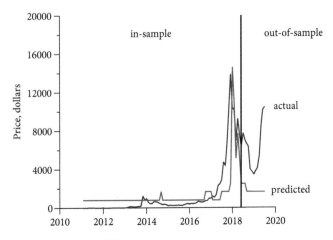

Figure 6.6 Jumanji and Bitcoin.

that correlation supersedes causation would not be troubled by the fact that Jumanji has nothing to do with Bitcoin. Correlation is enough.

In each case, using *Bitcoin* searches or *Jumanji* searches, there is a close correlation in-sample, during the rise and fall of Bitcoin prices, suggesting that a search term (*Bitcoin* or *Jumanji*) can be used to make profitable predictions of Bitcoin prices. Then the models completely whiff on the rebound in Bitcoin prices in 2019.

Correlations are easy to find. Useful correlations are more elusive.

Predicting Unemployment

The availability of data from Google, Twitter, Facebook, and other social media platforms provides an essentially unlimited number of variables than can be used for data mining—and an effectively unlimited number of spurious relationships that can be discovered.

To demonstrate this, we looked for some silly data that might be used to predict the U.S. unemployment rate. It didn't take long. The board game Settlers of Catan has five resources: brick, lumber, ore, sheep, and wool. Since unemployment data are monthly, we gathered monthly data from Google Trends on the frequency with which these five resources had been used in Google searches since January 2004—as far back as these data go.

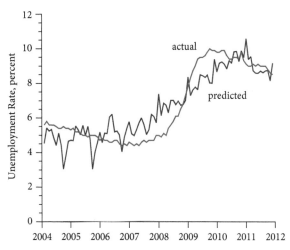

Figure 6.7 Using resources to predict unemployment.

We then estimated a model that used the number of monthly searches to predict the unemployment rate the next month from January 2004 through December 2011, leaving 2012 through 2019 for an out-of-sample test of the model.

Figure 6.7 shows that these five search terms did a pretty good job predicting the unemployment rate. The correlation between the actual unemployment rate and the predicted value, based on the previous month's searches, was an impressive 0.89. Should we apply for jobs as economic forecasters?

Remember that these are not real data on brick, lumber, ore, sheep, and wool—which might actually have something to do with the economy and the unemployment rate. These are the number of times people searched for these Settlers of Catan words on Google. Yet the 0.89 correlation is undeniable. Some data miners might think up possible explanations. Others might say that no explanation is necessary. Up is up.

Instead of thinking of explanations or deciding that explanations are not needed, let's see how our resource search model did predicting unemployment during the out-of-sample period, 2012 to 2019. Awful. Figure 6.8 shows that predicted unemployment fluctuated around nine percent, while actual unemployment fell below four percent.

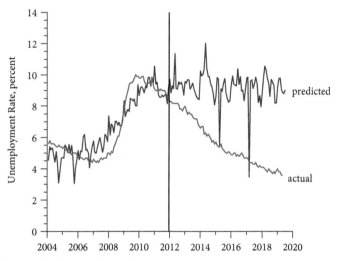

Figure 6.8 Resources flop out of sample.

There was no reason for Google searches for Settlers of Catan resources to have anything to do with the unemployment rate. They happened to be fortuitously correlated, but fortuitous correlations are temporary and worthless. This is a silly example of the broader point. In a world with limitless data, there is an endless supply of fortuitous correlations. It is not hard to find them, and it is difficult to resist being seduced by them—to being fooled by phantom patterns.

The Chinese Housing Bubble

We are not saying that Internet data are useless. We are saying that Internet data ought to be used judiciously. Researchers should collect relevant, reliable data with a well-defined purpose in mind, and not just pillage data looking for patterns. A recent study of the Chinese real estate market is a good example.

In China, all land belongs to the state, but in 1998 the Chinese State Council decided to privatize the property market by allowing local governments to sell land use rights lasting for several decades (typically seventy years) to property developers who construct buildings to sell to the public.

From 1998 to 2010, the annual volume of completed private housing units rose from 140 million square meters to over 610 million and homeownership rates reached 84.3 percent of the urban housing stock. In recent years, over a quarter of China's GDP has been tied to real estate construction.

One reason for this housing demand is that a considerable sex imbalance has existed in China for several decades, due largely to an ingrained cultural preference for male children and exacerbated by the one-child policy enacted from 1979 to 2015. Homeownership is desirable for men because it is a status signal that boosts their competitiveness in the marriage sweepstakes. Also, many Chinese parents help their children buy homes. Chinese households under the age of thirty-five have a fifty-five percent homeownership rate, compared to thirty-seven percent in the United States.

Figure 6.9 shows that Chinese residential real estate has been a better investment than stocks: housing has had a relatively attractive return with considerably less volatility. However, this is the kind of backward-looking, trend-chasing perspective that fuels speculative bubbles. People see prices rising and rush to buy, which causes prices to rise further—until they don't, which causes prices to collapse because there is no point in buying if prices aren't rising.

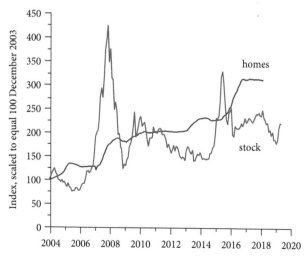

Figure 6.9 Shanghai stock market prices and prices of second-hand homes in Shanghai.

Some fear that the Chinese real estate market is a speculative bubble and point to the rapid growth in prices as evidence. From 2003 to 2013, China's first-tier cities—Beijing, Shanghai, Guangzhou, and Shenzhen—experienced an inflation-adjusted 13.1 percent annual increase in home prices. In China's second-tier and third-tier cities, the annual increases were 10.5 percent and 7.9 percent, respectively. Chinese billionaire and real estate magnate Wang Jianlin has warned that Chinese real estate is the "biggest bubble in history."

Gary and a student, Wesley Liang, investigated this question. A speculative bubble occurs when an asset's price rises far above the value of the income generated by the asset, because investors are not buying the asset for the income, but to sell for a higher price than they paid. In residential housing, the income is the rent saving (the cost of renting a comparable home), net of mortgage payments, property taxes, maintenance, and other expenses.

Gary and Wesley (who is fluent in Chinese) were able to obtain rent and price data from 链家 ("Lianjia"), one of China's largest real estate brokerage firms. They looked at residential apartment properties in China's two largest cities, Beijing and Shanghai, during the nine-month period from June 2018 to February 2019. The data include the rental price or sale price, location of the complex, number of rooms, square footage, facing direction, floor level, total number of floors in the building, and year built.

The Lianjia website requires a user to input a Chinese cellphone number on its website login page. A confirmation code is sent to the phone number and this code can be used on the website login page. Once logged in, past home sales and rental transaction data can be found under the section title "查交易", which translates to "check past transactions." The house sale and rental transactions were then manually matched.

There are two very important things about this study. First, it would have been essentially impossible for Gary and Wesley to collect these data without the Internet. Second, they used a theoretically valid model to determine which data to collect and analyze; they didn't surf the Internet, looking for variables that might happen to be correlated with Chinese home prices.

They were able to identify more than 400 matched pairs of homes in both Beijing and Shanghai. Table 6.2 shows that homes are typically small and expensive; however, to gauge whether the prices are justified by the rent savings, Gary and Wesley needed to compare what it would cost to buy and rent each matched pair.

Table 6.2 _Chinese housing data._

City	Average number of Rooms	Median Square Meters	Median Monthly Rent	Median Price	Median Price per Square Meter
Beijing	1.75	61.50	$954	$707,750	$11,517
Shanghai	1.64	55.59	$611	$409,750	$6,925

Table 6.3 _Annual rates of return, percent._

Holding Period	Beijing	Shanghai
One year	−15.10	−13.20
Ten years	−2.17	−0.92
Thirty years	1.11	1.70
Forever	2.50	3.16

Gary and Wesley initially assumed that buyers hold their homes forever, passing them from one generation to the next, not because homes are never resold, but because they wanted to calculate rates of return based solely on the income from the homes, without making any assumptions about future home prices. This is the same approach used by value investors to estimate stock market rates of return. Gary and Wesley also calculated rates of return over finite horizons using a variety of assumptions about how fast home prices might increase over the next one to thirty years.

Table 6.3 shows the results for an infinite horizon and for a finite horizon with three percent annual increases in home prices, rent, and expenses. Gary and Wesley used a variety of other assumptions to assess the robustness of the results.

The returns are often negative and clearly unattractive in comparison with Chinese ten-year government bonds that yield more than three percent and have substantially less risk. People who bought homes in Beijing or Shanghai as an investment in 2018 and 2019 were evidently anticipating that home prices would continue rising rapidly.

Home ownership in China may become even less financially attractive in the future. When the land use agreements expire, owners can apply for extensions, but there are no guarantees about the cost of obtaining an extension. One possibility is that the Chinese government will charge renewal fees based on current market prices. However, large lump-sum assessments are likely to be perceived as unfair and unaffordable for many homeowners. A less disruptive alternative would be to allow renewal fees to be paid in annual installments, akin to property taxes, or to impose explicit property taxes in lieu of renewal fees. Land-use renewal fees and/or property taxes will substantially reduce the already low income from home ownership.

These results indicate that the Beijing and Shanghai housing markets are in a bubble, and that home buyers should not anticipate continued double-digit annual increases in home prices. If double-digit price increases do continue, the Chinese real estate bubble will become even larger and more dangerous.

On the other hand, the possible consequences of a housing crash in China are so frightening that the Chinese government is unlikely to stand by and let it happen. The real estate market is too big to fail. If the air begins leaking out of the bubble, the Chinese government is likely to intervene through laws, regulations, or outright purchases to prevent a collapse. The Chinese real estate bubble will most likely end with a whimper, not a bang.

How to Avoid Being Misled by Phantom Patterns

The Internet provides a firehose of data that researchers can use to understand and predict people's behavior. However, unless A/B tests are used, these data are not from RCTs that allow us to test for causation and rule out confounding influences. In addition, the people using the Internet in general, and social media in particular, are surely unrepresentative, so data on their activities should be used cautiously for drawing conclusions about the general population.

Things we read or see on the Internet are not necessarily true. Things we do on the Internet are not necessarily informative. An unrestrained scrutiny of searches, updates, tweets, hashtags, images, videos, or captions is certain to turn up an essentially unlimited number of phantom patterns that are entirely coincidental, and completely worthless.

CHAPTER 7

The Reproducibility Crisis

In 2011, Daryl Bem published a now-famous paper titled "Feeling the Future." Bem reported the results of nine experiments he had conducted over a decade, eight of which showed statistically significant evidence of extrasensory perception (ESP). In one experiment, Cornell undergraduates were shown two curtains on a computer monitor—one on the left side and one on the right side—and asked to guess which curtain had an image behind it. Bem's novel twist on this guessing game was that the computer's random event generator chose the curtain that had an image after the student guessed the answer. When the image was an erotic picture, the students were able to guess correctly a statistically significant fifty-three percent of the time. The students felt the future. In another experiment, Bem found that people did better on a recall test if they studied the words they were tested on after taking the test. Again, the students felt the future.

Bem was a prominent social psychologist and his paper was published in a top-tier journal. Soon these remarkable studies were reported worldwide, including a front-page story in *The New York Times* and an interview on *The Colbert Report*. Bem invited others to replicate his results, and even provided step-by-step instructions on how to redo his experiments.

Bem was so well-known, the journal so well-respected, and the results so unbelievable, that several researchers took up the challenge and attempted to replicate Bem's experiments. They found no evidence that people could feel the future. However, Bem's paper was not in vain. It was one of two events that made 2011 a watershed year for the field of social psychology.

Across the Atlantic, forty-five-year-old Danish social psychologist Diederik Stapel had become a dean at Tilburg University and made a name for himself with the publication of dozens of papers with catchy names like:

One of his most well-known papers reported that eating meat made people selfish and anti-social; another reported that messy environments made people racist. An unpublished paper found that people filling out a questionnaire were more likely to eat M&Ms from a nearby coffee cup if the cup had the word "kapitalisme" on it, rather than a meaningless jumble of the same letters.

Unlike Bem, many of Stapel's papers were plausible. Unlike Bem, many of Stapel's data were fake.

In 2011 Tilburg University suspended Stapel for fabricating data; fifty-eight of his published papers have now been retracted—including the papers mentioned above.

Stapel later explained that he "wanted too much, too fast," but he was also driven by "a quest for aesthetics, for beauty—instead of the truth." He preferred orderliness and clear-cut answers. He wanted patterns so badly that he made them up when he couldn't find them.

At the beginning of his fabrication journey, he collected data and cleaned the data up afterward—changing a few numbers here and there—until he got the answers he wanted. Later, he didn't bother collecting data. He just made up reasonable numbers that supported his theories. In the M&M study, he sat near a cup full of M&Ms and ate what seemed to be a reasonable number; then he pretended that he had done an experiment with several volunteers and simply made up numbers that were consistent with his own M&M consumption.

There are two very interesting things about the Bem and Stapel 2011 incidents. First, they are not isolated and, second, they are not that different.

Sloppy Science

The three universities at which Stapel had worked (Tilburg University, the University of Groningen, and the University of Amsterdam) convened inquiry committees to investigate the extent of Stapel's wrongdoings. Their final report was titled, "Flawed Science: The Fraudulent Research Practices of Social Psychologist Diederik Stapel." The key part of the title is the first two words: "Flawed Science." The problems they found were larger than Stapel's fabrications. They were at the heart of the way research is often done.

The committees were surprised and disappointed to find widespread "sloppy science":

A "byproduct" of the Committees' inquiries is the conclusion that, far more than was originally assumed, there are certain aspects of the discipline itself that should be deemed undesirable or even incorrect from the perspective of academic standards and scientific integrity.

Another clear sign is that when interviewed, several co-authors who did perform the analyses themselves, and were not all from Stapel's "school", defended the serious and less serious violations of proper scientific method with the words: that is what I have learned in practice; everyone in my research environment does the same, and so does everyone we talk to at international conferences.

The report includes a long section on *verification bias*, massaging the data in order to obtain results that verify the researcher's desired conclusion:

One of the most fundamental rules of scientific research is that an investigation must be designed in such a way that facts that might refute the research hypotheses are given at least an equal chance of emerging as do facts that confirm the research hypotheses. Violations of this fundamental rule … essentially render the hypotheses immune to the facts.

They give several examples:

1 An experiment is repeated (perhaps with minor changes) until statistically significant results are obtained; this is the only experiment reported.
2 Several different tests are done, with only the statistically significant results reported.
3 The results of several experiments are combined if this merging produces statistically significant results.
4 Some data are discarded in order to produce statistically significant results, either with no mention in the study or with a flimsy ad hoc excuse, like "the students just answered whatever came into their heads."

5 Outliers in the data are retained if these are needed for statistical significance, with no mention of the fact that the results hinge on a small number of anomalous data.

When the committees had access to complete data sets and found such practices in published papers, the researchers were quick to offer excuses. Even more disheartening, the researchers often considered their actions completely normal.

The committees also found that leading journals not only ignored such mischief, but encouraged it:

Reviewers have also requested that not all executed analyses be reported, for example by simply leaving unmentioned any conditions for which no effects had been found, although effects were originally expected. Sometimes reviewers insisted on retrospective pilot studies, which were then reported as having been performed in advance. In this way the experiments and choices of items are justified with the benefit of hindsight.

Not infrequently reviews were strongly in favour of telling an interesting, elegant, concise and compelling story, possibly at the expense of the necessary scientific diligence.

The conclusions of the committees investigating Stapel implied that, other than proposing an absurdity, there was nothing unusual about Bem's paper. It was what most social psychology researchers did. Bem's friend Lee Ross said that, "The level of proof here was ordinary. I mean that positively as well as negatively. I mean it was exactly the kind of conventional psychology analysis that [one often sees], with the same failings and concerns that most research has."

Stapel and Bem are not all that different. They are both on a continuum that ranges from minor data tweaking to wholesale massaging to complete fabrication. Is running experiments until you get what you want that much different from taking the short cut of making up data to get the results you want?

The title of an article in *Slate* magazine stated the problem succinctly: "Daryl Bem Proved ESP Is Real—Which Means Science is Broken." The credibility of scientific research has been undermined by what has come to be called the *reproducibility crisis* (or the replication crisis), in that attempts to replicate published studies often fail; in Stapel's case, because he made up data; in Bem's case, because he manipulated data.

In the erotic-picture study, for example, Bem did his experiment with five different kinds of pictures, and chose to emphasize the only kind that was statistically significant. Statistician Andrew Gelman calls this "the garden of forking paths." If you wander through a garden making random choices every time you come to a fork in the road, your final destination will seem almost magical. What are the chances that you would come to this very spot? Yet, you had to end up somewhere. If the final destination had been specified before you started your walk, it would have been amazing that you found your way to it. However, identifying your destination after you finish your walk is distinctly not amazing.

To illustrate the consequences of verification bias, three researchers had twenty undergraduates at University of Pennsylvania listen to the Beatles' song, "When I'm 64." The researchers then used some sloppy science to demonstrate that the students were one and a half years younger after they had listened to the song. They didn't mention Bem, but the similarities were clear.

During one interview, Bem said, without remorse,

I would start one [experiment], and if it just wasn't going anywhere, I would abandon it and restart it with changes. I didn't keep very close track of which ones I had discarded and which ones I hadn't. I was probably very sloppy at the beginning. I think probably some of the criticism could well be valid. I was never dishonest, but on the other hand, the critics were correct.

In a September 2012 e-mail to social psychology researchers, Nobel laureate Daniel Kahneman warned that, "I see a train wreck looming." Gary was at a conference a few years later when a prominent social psychologist said that his field was the poster child for irreproducible research and added that, "My default assumption is that anything published in my field is wrong."

The reproducibility crisis had begun.

Ironically, some now wonder if Bem's mischief was deliberate. Had he spent ten years preparing an elaborate prank that would be a wake-up call to others in the field? Did he publish preposterous results and encourage others to try to replicate them in order to clean up social psychology? Bem's own words in a 2017 interview with *Slate* magazine suggest the answer is no, but perhaps this is part of the prank:

I'm all for rigor, but I prefer other people do it. I see its importance—it's fun for some people—but I don't have the patience for it…If you looked at all my past experiments, they were always rhetorical devices. I gathered data to show how my point would be made. I used data as a point of persuasion, and I never really worried about, "Will this replicate or will this not?"

Bem and Stapel happened to be social psychologists, but the replication crisis extends far beyond social psychology. In medical research, many "proven-effective" medical treatments are less effective in practice than they were in the experimental tests. This pattern is so common, it even has a name—the "decline effect."

For example, Reserpine was a popular treatment for hypertension until researchers reported in 1974 that it substantially increased the chances of breast cancer. Its use dropped precipitously and it is seldom used today. However, several attempts to replicate the original results concluded that the association between Reserpine and breast cancer was spurious. One of the original researchers later described the initial study as the "Reserpine/ breast cancer disaster."

In retrospect, he recognized that:

We had carried out, quite literally, thousands of comparisons involving hundreds of outcomes and hundreds (if not thousands) of exposures. As a matter of probability theory, "statistically significant" associations were bound to pop up and what we had described as a possibly causal association was really a chance finding.

They had walked through a garden of forking paths.

A 2018 survey of 390 professional statisticians who do statistical consulting for medical researchers found that more than half had been asked to do things they considered severe violations of good statistical practice, including conducting multiple tests after examining the data and misrepresenting after-the-fact tests as theories that had been conceived before looking at the data. A 2011 survey of more than 2,000 research psychologists found that most admitted to having used questionable research practices. A 2015 survey by *Nature*, one of the very best scientific journals, found that more than seventy percent of the people surveyed reported that they had tried and failed to reproduce another scientist's experiment and more than half had tried and failed to reproduce some of their own studies!

The Reproducibility Project launched by Brian Nosek looked at 100 studies reported in three leading psychology journals in 2008. Ninety-seven of these studies reported significant results which, itself, suggests a problem. Surely, ninety-seven percent of all well-conducted studies do not yield significant results. Two hundred and seventy researchers volunteered to attempt to replicate these ninety-seven studies. Only thirty-five of the ninety-seven original conclusions were confirmed and, even then, the effects were invariably smaller than originally reported.

The Experimental Economics Replication Project attempted to replicate eighteen experimental economics studies reported in two leading economics journals during the years 2011–2014. Only eleven (sixty-one percent) of the follow-up studies found significant effects in the same direction as originally reported.

One reason for the reproducibility crisis is the increased reliance on data-mining algorithms to build models unguided by human expertise. Results reported with data-mined models are inherently not reproducible, since they will almost certainly include patterns in the data that disappear when they are tested with new data.

You Are What Your Mother Eats

In 2008, a team of British researchers looked at 133 food items consumed by 740 British women prior to pregnancies and concluded that women who consumed at least one bowl of breakfast cereal daily were much more likely to have male babies when compared with women who consumed one bowlful or less per week. After this remarkable finding was published in the prestigious Proceedings of the Royal Society with the catchy title "You Are What Your Mother Eats," it was reported worldwide and garnered more than 50,000 Google hits the first week after it was published. There was surely an increase in cereal consumption by women who wanted male babies and a decrease in cereal consumption by women who wanted female babies.

The odd thing about this conclusion, as quickly pointed out by another researcher, Stanley Young, is that a baby's sex is determined by whether the male sperm is carrying an X or Y chromosome: "The female has nothing to do with the gender of the child." So, how did the research team find a statistically significant relationship between cereal and gender? They looked at the consumption of 132 food items at three points in time (before conception, sixteen weeks after conception, and sixteen to twenty-eight weeks after), giving a total of 396 variables. In addition, Young and two co-authors wrote that "there also seems to be hidden multiple testing as many additional tests were computed and reported in other papers." It is hardly surprising that hundreds of tests yielded a statistically significant relationship for one food item at one point in time (cereal before conception).

Zombie Studies

Jay received the following email from a friend (who coincidentally had been a student of Gary's):

The risks and efficacy of vaccines … is a particularly hot topic here in Oregon as the state legislature is working on a bill that removes the option for a non-medical exemption from vaccination for school children. If the child is not vaccinated and does not have a medical exemption, that child will not be allowed to attend public school.

I have been told that there are studies that support both conclusions: vaccines do cause autism and other auto-immune disease and vaccines do not cause these conditions. I have not done any research myself. I understand that the linchpin study supporting the harmfulness of vaccines has been retracted. What is the truth?

I have friends looking to move out of the state if this bill becomes law.

I would like to understand the science before addressing the personal liberty issue of such a law.

The 1998 study that claimed to have found an association between MMR vaccines and autism was indeed retracted for a variety of statistical sins, twelve long years after it was published. (The details are in Gary's book, *Standard Deviations: Flawed Assumptions, Tortured Data, and Other Ways to Lie with Statistics*.) While there is always a slight risk of an allergic reaction, there is no evidence of a link between MMR vaccines and autism, and the dangers they prevent are far greater than any risks they create. Unfortunately, like most retracted studies, this one has had a life after retraction. The paper was retracted in 2010, with the journal's editor-in-chief stating that the paper was "utterly false." The researcher was also barred from practicing medicine in England. Yet, some people still believe the paper's bogus claims. It has become a zombie study.

Many retracted papers, including some of Stapel's papers, are still being cited by researchers who are blissfully unaware of their demise. So are papers that have been discredited, though not retracted. Many websites continue to report that it has been scientifically proven that eating cereal before conception can increase a woman's chances of having a boy baby and that abstaining from cereal can increase the chances of a girl baby.

Another example is a now-discredited claim that hurricanes with female names are deadlier than hurricanes with male names (full disclosure: Gary was one of the debunkers.) The hurricane data were not made up, but the paper committed many of the verification-bias sins listed above. For example, the paper included pre-1979 hurricanes when all hurricanes had female names and hurricanes were deadlier because of weaker buildings and less advance warning. During the post-1979 years, when female and male names alternated, there is no difference in deadliness. Yet, as the hurricane season approaches every year, social media are reliably abuzz with dire warnings about hurricanes with female names.

The Reproducibility Crisis: A Case Study

In Chapter 6, we discussed a 2018 study by Aleh Tsyvinski and Yukun Liu that asserted that Bitcoin prices could be predicted from how often the word *Bitcoin* is mentioned in Google searches. They also looked at correlations between Bitcoin prices and hundreds of other economic variables and concluded that Bitcoin "represents an asset class that can be assessed using simple finance tools."

Their study was essentially unrestrained data mining because there is no logical reason for Bitcoin prices to be related to anything other than investor sentiment. Unlike bonds that yield interest, stocks that yield dividends, apartments that yield rent, businesses that yield profits, and other real investments, Bitcoin doesn't yield anything at all, so there is no compelling way to value Bitcoin the way investors can value bonds, stocks, apartments, businesses, and other real investments.

Bitcoin prices can fluctuate wildly because people buy Bitcoins when they expect the price to go up and sell when they expect the price to go down—which causes Bitcoin prices to go up and down even more. Such self-fulfilling expectations need not be related to anything that is measurable. They may be driven by little more than what the great British economist John Maynard Keynes called "animal spirits."

Attempts to relate Bitcoin prices to real economic variables are destined to disappoint, yet Liu and Tsyvinski tried to do exactly that, and their disappointments are instructive. To their credit, unlike many studies, they did not hide their failed results; they were transparent about how many variables they considered: 810 of them! When considering so many variables, it is strongly advised to use a much stricter standard than the usual five-percent hurdle for statistical significance (as an extreme example, particle physicists only consider data to be compelling if the likelihood that it would happen by chance is less than once in 3.5 million). Liu and Tsyvinski adjusted the hurdle, but in the wrong direction, and considered any association with less than a ten percent chance of occurring by luck to be statistically significant!

Liu and Tsyvinski used data from January 1, 2011 through May 31, 2018. For our replication tests, we use the out-of-sample period from June 1, 2018 through July 31, 2019. Their variables include the Canadian dollar–U.S. dollar exchange rate, the price of crude oil, stock returns in the healthcare industry, and stock returns in the beer industry. The occasional justifications they offer are seldom persuasive.

For example, Liu and Tsyvinski acknowledge that, unlike stocks, Bitcoins don't generate cash or pay dividends, so they used what they call a "similar metric," the number of Bitcoin Wallet users:

Obviously, there is no direct measure of dividend for the cryptocurrencies. However, in its essence, the price-to-dividend ratio is a measure of the gap between the market value and the fundamental value of an asset. The market value of cryptocurrency is just the observed price. We proxy the fundamental value by using the number of Bitcoin Wallet users.

The number of Bitcoin Wallet users is not a substitute for cash dividends paid to stockholders. This farfetched proxy is reminiscent of the useless metrics (like website visitors) that wishful investors conjured up during the dot-com bubble to justify ever higher stock prices.

One plausible relationship that they did consider is between current and past Bitcoin returns. Figure 7.1 shows Bitcoin prices during the period they studied. Figure 7.2 shows the volume of trading during this same period. The result is, in essence, a speculative bubble with persistently rising prices persuading speculators to splurge—putting further upward pressure on prices. Figure 7.1 and Figure 7.2 are strongly consistent with the idea that speculators rushed to buy Bitcoin as the price went up, and fled afterward.

To the extent that Bitcoin was a speculative bubble for much of the time period studied by Liu and Tsyvinski, it is reasonable to expect

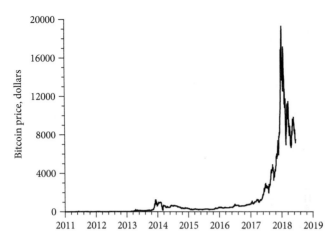

Figure 7.1 The Bitcoin bubble.

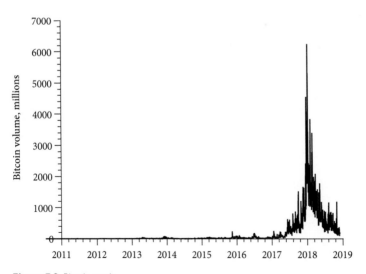

Figure 7.2 Bitcoin mania.

strong momentum in prices. However, once the bubble pops, momentum will evaporate and no longer be very useful for predicting Bitcoin price movements.

As discussed in Chapter 6, Liu and Tsyvinski consider how well Google searches for the word *Bitcoin* predict Bitcoin prices. They also reverse the direction of the relationship in order to see how well Bitcoin prices predict weekly Google searches for the word *Bitcoin*. The effect of Bitcoin prices on *Bitcoin* searches is not particularly useful, but at least there is a plausible explanation for such a relationship—unlike the vast majority of variables they analyzed.

Count Them Up

Table 7.1 summarizes the results. Overall, Liu and Tsyvinski estimated 810 correlations between Bitcoin prices and various variables, and found sixty-three relationships that were statistically significant at the ten percent level. This is somewhat fewer than the eighty-one statistically significant relationships that would be expected if they had just correlated Bitcoin prices with random numbers.

Seven of these sixty-three correlations continued to have the same signs and be statistically significant out-of-sample. Five of these seven correlations were for equations using Bitcoin prices to predict *Bitcoin* searches, which are arguably the most logically plausible relationships they consider. Ironically, this finding is an argument against the data mining they did and in favor of restricting one's attention to logical relationships—these are the ones that endure.

For the hundreds of other relationships they consider, fewer than ten percent were significant in-sample at the ten percent level, and fewer than ten percent of these ten percent continued to be significant out-of-sample. Of course, with enough data, coincidental patterns can always be found

Table 7.1 *Bitcoin estimated coefficients.*

	Number
Estimated coefficients	810
Coefficients significant in-sample	63 of 810
Coefficients significant (and with same signs) in-sample and out-of-sample	7 of 63

and, by luck alone, some of these coincidental patterns will persist out of sample. Should we conclude that, because Bitcoin returns happened to have had a statistically significant negative effect on stock returns in the paperboard-containers-and-boxes industry that was confirmed with out-of-sample data, a useful, meaningful relationship has been discovered? Or should we conclude that these findings are what might have been expected if all of the estimated equations had used random numbers with random labels instead of real economic variables?

The authors don't attempt to explain the relationships that they found: "We don't give explanations; we just document this behavior." Patterns without explanations are treacherous. A search for patterns in large databases will almost certainly result in their discovery, and the discovered patterns are likely to vanish when the results are used to make predictions. What is the point of documenting patterns that vanish?

The Charney Report

A research study concluded that, "We now have incontrovertible evidence that the atmosphere is indeed changing and that we ourselves contribute to that change. Atmospheric concentrations of carbon dioxide are steadily increasing, and these changes are linked with man's use of fossil fuels and exploitation of the land." You could be forgiven for assuming this is a recent scientific response to climate-change deniers, but you would be mistaken; the study was published in 1979.

The report was titled "Carbon Dioxide and Climate: A Scientific Assessment" and was produced for the National Academy of Sciences by a study group led by Jule Charney. It made an alarming prediction: if the concentration of CO_2 were to double, "the more realistic of the modeling efforts predict a global surface warming of between 1.5 degrees Centigrade and 4.5 degrees Centigrade, with greater increases at high latitudes." The report also made the disturbing prediction that warming may be delayed by the heat-trapping capability of the ocean, but reaching the predicted thermal equilibrium temperature would be inevitable. In other words, even when it becomes clear that global warming is a major problem and humanity becomes carbon neutral, the warming will continue for decades.

Skepticism was understandable forty years ago when it wasn't self-evident that the world was experiencing a warming trend. There were even a few articles at the time warning of "global cooling." Nonetheless, most climate

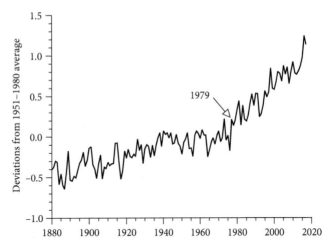

Figure 7.3 Calling a trend before it happens.

scientists found the evidence compelling. Today, forty years later, we can check in on their prediction and see how it fared. Figure 7.3 shows annual global surface temperatures back to 1880, as far back as these NASA data go, and identifies the year this study was published. The data are shown as deviations from what the average temperature was for the thirty-year period, 1951 through 1980.

Since 1979, CO_2 has increased by twenty-one percent and the global average surface temperature has gone up by 0.66° C, which is on track for a 3° C increase in global surface temperatures accompanying a 100 percent increase in CO_2—right in the middle of the range predicted by the Charney Report. How did these scientists produce a report that made a reliable forty-year prediction?

The first thing working in the group's favor was that it was led by an expert with few peers, either technically or in terms of scientific integrity. In fact, if you had to pick one man to thank for laying the foundation for the modern-day successes of weather prediction, you would be likely to choose Jule Charney. Born in San Francisco on New Year's Day in 1917, Charney completed his doctorate at UCLA in 1946. He then worked on the earliest weather models at Princeton in 1950, where his equations revolutionized the field of numeric weather prediction (including how to predict cyclones). He went on to mentor countless students while continuing to produce groundbreaking research during twenty-five years at MIT.

The next thing the group did right was come up with a hypothesis and prediction based in physics and chemistry, rather than search for (possibly spurious) patterns in the data. It is tempting to find a trend and make up a story for why it will continue, but serious science goes the other way around: theory first, and data later.

The final ingredient that contributed to the success of this timeless paper was a healthy dose of "not fooling themselves." They were careful not to overstate what they could predict confidently, and they considered every reason that they could think of that might make their predictions wrong. Unfortunately, Charney died in 1981 and did not live to see the validation of his team's predictions.

Skepticism vs Denialism

The fact that research papers are sometimes sloppy science does not mean that the consensus conclusions of science should be dismissed. The reproducibility crisis encourages skepticism, but not denialism.

We should be skeptical of individual studies that are flawed in various ways, but we should also be skeptical of claims that exaggerate scientific uncertainty, or dismiss entire fields of research in order to further a political or economic agenda. In the case of climate change, there are approximately 200 worldwide scientific organizations that support the view that climate change is caused by human action and none that reject it. (Not surprisingly, the American Association of Petroleum Geologists was the last holdout, but switched from opposing the consensus to being non-committal in 2007.)

Of course, science is not an opinion poll. A 1931 book that was skeptical of Albert Einstein's theory of relativity was titled, *100 Authors against Einstein*. Einstein's response was, "Why 100? If I were wrong, one would have been enough." The case for global warming is not based on a survey, but on a convergence of evidence from multiple lines of inquiry—pollen, tree rings, ice cores, coral reefs, glacial and polar ice-cap melt, sea-level rise, ecological shifts, carbon dioxide increases, the unprecedented rate of temperature increase—that all converge to one conclusion.

The reproducibility crisis is real, but it doesn't affect all fields equally and should not be used as fodder for political purposes. We all benefit if science remains fruitful, productive, and insulated from ideologues. Be wary of groups that exaggerate uncertainties in order to promote their own self interests. The tobacco industry wrote the playbook on this and climate change deniers are following it today.

The scientific community is responding to the reproducibility crisis by establishing procedures to identify and thwart the various sins identified by the committees investigating Diederik Stapel. Many journals now require that research plans be filed before the research begins and insist that all data used in the research be made publicly available. The scientific enterprise is one human endeavor where long-term progress is indisputable and it needs to stay that way.

How to Avoid Being Misled by Phantom Patterns

Attempts to replicate reported studies often fail because the research relied on data mining—searching through data for patterns without any pre-specified, coherent theories. The perils of data mining can be exacerbated by data torturing—slicing, dicing, and otherwise mangling data to create patterns. If there is no underlying reason for a pattern, it is likely to disappear when someone attempts to replicate the study.

Big data and powerful computers are part of the problem, not the solution, in that they can easily identify an essentially unlimited number of phantom patterns and relationships, which vanish when confronted with fresh data.

If a researcher will benefit from a claim, it is likely to be biased. If a claim sounds implausible, it is probably misleading. If the statistical evidence sounds too good to be true, it probably is.

CHAPTER 8

Who Stepped in It?

During Sweden's Age of Greatness (1611–1718), a series of ambitious kings transformed a rural backwater into a military powerhouse. Swedish firearms made of copper and iron were exceptional. Swedish ships made of virgin hardwood were fearsome. Swedish soldiers were heroic. Sweden ruled the Baltic.

In 1625 King Gustav II placed orders for four military ships from the Hybertsson shipyards in Stockholm. At the time there was little theory to guide shipbuilding. Designers did not know how to determine the center of gravity or the stability of various designs. It was literally trial-and-error—learning what worked and didn't work by seeing which ships were seaworthy and which toppled over and sank. Henrik Hybertsson was the master shipwright at the Hybertsson shipyards and he started construction of the Vasa, a traditional 108-foot ship, for the King without detailed plans or even a rough sketch, since he and the builders were very familiar with such ships.

Most military ships at the time carried hundreds of soldiers and had a single gun deck armed with small cannons. The conventional military strategy was to use the cannons to disable an enemy ship, which could then be boarded by soldiers. King Gustav favored an alternative strategy of using cannon fire to sink enemy ships, and he changed the order for the Vasa to a 120-foot ship armed with 3,000-pound cannons firing twenty-four-pound cannon balls.

When he learned that Denmark was building a ship with two gun decks, the King ordered that a second gun deck be added to the Vasa, and that the length of the boat's keel be extended to 135 feet. In addition to the extra cannons, a high gun deck would allow the cannon balls to travel farther. A 111-foot keel had already been built, so Hybertsson scaled the design upward—still working without detailed plans. Hybertsson had never built a ship with two gun decks, but it was assumed to be a simple extrapolation of a 108-foot ship with a single gun deck.

The 111-foot keel consisted of three pieces of oak joined together end to end. A fourth piece of oak was added to one end to get the Vasa's keel up to 135 feet. One problem was that the width and depth of the keel should have been increased too, but it was too late to do so, which made the boat less stable. In addition, to accommodate the second gun deck, the ship was widened at the top, but could not be widened at the bottom, again making the boat less stable, and there was limited room for ballast which might have lowered the ship's center of gravity. Even if they had been able to add enough ballast to lower the center of gravity sufficiently, it would have pulled the first gun deck below sea level.

Instead of thirty-two guns on one deck, Vasa carried forty-eight guns, twenty-four on each deck, which raised the ship's center of gravity. The King also ordered the addition of hundreds of gilded and painted oak carvings high on the ship, where enemy soldiers could see them; these, too, raised the center of gravity. You know where this story (and this ship) is heading.

A final empirical mishap was that some worker used rulers calibrated in Swedish feet, which are divided into twelve inches, and other workers used rulers calibrated in Amsterdam feet, which are divided into eleven inches. So, six inches meant different things to different workers. The result was that the Vasa was lopsided—heavier on the port side than on the starboard side.

When the Vasa was launched in Stockholm harbor on August 10, 1628, the wind was so light that the crew had to pull the sails out by hand, hoping to catch enough of a breeze to propel the boat forward. Twenty minutes later, 1,300 feet from shore, a sudden eight-knot gust of wind toppled the boat and it sank.

The wood survived underwater for hundreds of years because of the icy, oxygen-poor water in the Baltic Sea, and 333 years later, in 1961, the Vasa was pulled from the ocean floor. After being treated with a preservative for seventeen years, it is now displayed in its own museum in Stockholm.

The Vasa is a tragic example of the importance of theory and the perils of relying on patterns, as Hybertsson did when he took a pattern that worked for 108-foot boats and tried to apply it to a 135-foot boat. We now know a lot about boat stability and we know that taking a boat design that works for one boat size and scaling up, say, by increasing all of the dimensions by twenty-five percent, can end disastrously.

The Vasa is not an anomaly. Being fooled by phantom patterns has been going on for a very long time and caused a great many awkward blunders and pratfalls.

New Coke

Thousands of charlatans have peddled thousands of patent medicines, so-named because their promoters claim that a secret formula has been proven effective and patented. Usually, neither is true. In order to patent a medicine, the secret formula would have to be revealed. In order to be proven effective, the medicine would have to be tested. Most elixirs were never tested, only promoted.

Patent medicines were made from exotic herbs, spices, and other ingredients, and were said to cure most anything, including indigestion, tuberculosis, and cancer. Cure-all nostrums containing snake oil were the source of today's insult: *snake-oil salesman*. Alcohol was often a primary ingredient. Sometimes, there was morphine, opium, or cocaine.

In the second half of the nineteenth century, a wildly popular concoction was Vin Mariani, a mixture of Bordeaux wine and coca leaves created by a French chemist named Angelo Mariani. The wine extracted cocaine from the coca leaves, making for a potent combination of wine and cocaine. Queen Victoria, Thomas Edison, and thousands of prominent performers, politicians, and Popes consumed Vin Mariani.

Colorful posters and advertisements claimed that Vin Mariani "fortifies and refreshes body & brain. Restores health and vitality."

Pope Leo XIII said that he drank Vin Mariani to "fortify himself when prayer was insufficient." He was so enamored of this coca wine that he awarded it a special medal, which Mariani quickly turned into an advertising opportunity.

In the United States, a former Confederate Colonel named John Pemberton had been wounded in the Civil War and was addicted to morphine. He ran a drug store (Pemberton's Eagle Drug and Chemical House) in Georgia and, in 1885, developed a nerve tonic he called Pemberton's French Wine Coca, which he evidently patterned after Vin Mariani and intended not only to substitute for morphine but enjoy some of the commercial success of Vin Mariani. In addition to wine and coca, he added kola nuts, which provided caffeine. In 1886 local authorities prohibited the making, selling, or consuming of alcoholic beverages. So, Pemberton replaced the wine with carbonated water and sugar, and called his de-wined tonic Coca-Cola, reflecting the stimulating combination of coca leaves and kola nuts.

Pemberton advertised Coca-Cola as "the temperance drink" that would cure many diseases, including headaches, morphine addiction, and, as a bonus, was "a most wonderful invigorator of sexual organs."

Sales were modest and, in 1888, another druggist, Asa Griggs Candler, acquired the Coca-Cola recipe and marketing rights, reportedly for $2,300. Candler launched the Coca-Cola Company, which went on to dominate the world soft-drink market. One of his innovations was to sell Coca-Cola in bottles. This allowed people to drink Coke outside, including blacks, who were not permitted to sit at soda fountains in the South.

Candler made a few changes to the original formula in order to improve the taste and he also responded to a public backlash against cocaine (that culminated in cocaine being made illegal in the United States in 1914) by removing cocaine from the company's beverages. In 1929, Coke scientists figured out how to make a coca extract that is completely cocaine-free. The Drug Enforcement Administration now allows Coca-Cola to import coca leaves to a guarded chemical processing facility in New Jersey that produces the cocaine-free fluid extract, labeled "Merchandise No. 5."

The current formula for Coca-Cola is a closely guarded secret, kept in a vault and said to be known by only two company employees, who are not allowed to travel together and whose identities are also a secret.

In 1985, after ninety-nine years with essentially the same taste, Coca-Cola decided to switch from sugar to a new, high-fructose corn syrup, to make Coke taste sweeter and smoother—more like its arch rival, Pepsi. This historic decision was preceded by a top-secret $4 million survey of 190,000 people, in which the new formula beat the old by fifty-five percent to forty-five percent. What Coke's executives neglected to take into account was that many of the forty-five percent who preferred old Coke did so passionately. The fifty-five percent who voted for New Coke might have been able to live with the old formula, but many on the other side swore that they could not stomach New Coke.

Coca-Cola's announced change provoked outraged protests and panic stockpiling by old-Coke fans. Soon, Coca-Cola backed down and brought back old Coke as "Coke Classic." A few cynics suggested that Coca-Cola had planned the whole scenario as a clever way of getting some free publicity and creating, in the words of a senior vice-president for marketing, "a tremendous bonding with our public." In 1985, New Coke captured 15.0 percent of the entire soft-drink market and Coke Classic 5.9 percent with Pepsi at 18.6 percent. In 1986, New Coke collapsed to 2.3 percent, Coke Classic surged to 18.9 percent, and Pepsi held firm at 18.5 percent.

In 1987, *The Wall Street Journal* commissioned a survey of 100 randomly selected cola drinkers, of whom fifty-two declared themselves beforehand to be Pepsi partisans, forty-six Coke Classic loyalists, and two New Coke drinkers. In the *Journal*'s blind taste test, New Coke was the winner with forty-one votes, followed by Pepsi with thirty-nine, and Coke Classic with twenty. Seventy of the 100 people who participated in the taste tests mistakenly thought they had chosen their favorite brand; some were indignant. A Coke Classic drinker who chose Pepsi said, "I won't lower myself to drink Pepsi. It is too preppy. Too yup. The New Generation—it sounds like Nazi breeding. Coke is more laid back." A Pepsi enthusiast who chose Coke said, "I relate Coke with people who just go along with the status quo. I think Pepsi is a little more rebellious and I have a little bit of rebellion in me."

In 1990, Coca-Cola relaunched New Coke with a new name—Coke II—and a new can with some blue color, Pepsi's traditional color. It was no more successful than New Coke.

Coca-Cola executives and many others in the soft-drink industry remain convinced that cola drinkers prefer the taste of New Coke, even while they remain fiercely loyal to old Coke and Pepsi—a loyalty due more to advertising campaigns than taste. Given the billions of dollars that cola companies spend persuading consumers that a cola's image is an important part of the taste experience, blind taste tests may simply be irrelevant.

Data that are riddled with errors and omissions, or are simply irrelevant, are why "garbage in, garbage out" is such a well-known aphorism. In 2016, IBM estimated that the annual cost of decisions based on bad data was $3.1 trillion. Being duped by data is a corollary of being fooled by phantom patterns.

Telling Bill Clinton the Truth

Two weeks before the 1994 congressional elections, *The New Yorker* published an article about the Democrat Party's chances of keeping control of Congress. President Bill Clinton's top advisors were confident that Republicans were unpopular with voters—a confidence based on extensive surveys showing strong disapproval. But the article's author observed that:

The Democratic pollsters had framed key questions in ways bound to produce answers the President presumably wanted to hear. When the pollsters asked in simple unadorned, neutral language about the essential ideas in the Republican agenda—lower taxes, a balanced budget, term limits, stronger defense, etc.— respondents approved in large numbers.

One of the president's pollsters explained how she had to "frame the question very powerfully" in order to get the desired answers. Thus, one of the questions asked by the president's pollsters was:

[Republican candidate X] signed on to the Republican contract with their national leadership in Washington saying that he would support more tax cuts for the wealthy and increase defense spending, paid for by cuts in Social Security and Medicare. That's the type of failed policies we saw for twelve years, helping the wealthy and special interests at the expense of the middle class. Do you approve?

When the writer at *The New Yorker* suggested to a top Clinton advisor that such loaded questions gave the president a distorted picture of voter opinion, he responded, "That's what polling is all about, isn't it?" *The New Yorker* also quoted an unidentified "top political strategist" for the Democrats who was not part of Clinton's inner circle:

The President and his political people do not understand what has happened here. Not one of them ever comes out of that compound. They get in there at 7 A.M. and leave at 10 P.M., and never get out. They live in a cocoon, in their own private Disney World. They walk around the place, all pale and haggard, clutching their papers, running from meeting to meeting, and they don't have a clue what's going on out here. I mean, not a clue.

The election confirmed this strategist's insight and refuted the biased polls. The Republicans gained eight seats in the U.S. Senate and fifty-six seats in the House of Representatives, winning control of Congress for the first time in forty years. The Republicans also gained twelve governorships, giving them a total of thirty (including seven of the eight largest states), and gained more than 400 seats in the state legislatures, giving them majorities in seventeen states formerly controlled by Democrats.

When the President held a press conference the day after the election, a columnist at *The Washington Post* wrote that Clinton was "pretty much in the Ancient Mariner mode, haunted and babbling." *The New Yorker* reported that, "It was a painful thing to watch…[The] protestations of amity and apology were undercut by the President's over-all tone of uncomprehending disbelief."

Contrary to the opinion of Clinton's advisors, polling is not all about asking loaded questions. Clinton would have been much better served by fairly worded surveys that had given him and his advisors a clear idea of what voters wanted. After the 1994 debacle, Clinton replaced most of his pollsters and, once he had better information about voters' concerns, won a landslide re-election in 1996. Bill Clinton was one of the greatest campaigners in U.S. history, but he needed good data.

Stock Market Secrets

The stock market has been an enduring inspiration for people who look for patterns in data. There are *lots* of data and a useful pattern can make *lots* of money.

The problem is that stock prices are not determined by physical laws, like the relationship between the volume and pressure of a gas at a given temperature. Stock prices are determined by the willingness of investors to buy and sell stock.

If there is good news about a company (for example, an increase in its earnings), the stock price will rise until there is a balance between buyers and sellers. This makes sense, but the thing about news is that, by definition, it is unpredictable. If investors know something is going to happen, it won't be news when it does happen. For instance, during the 1988 presidential election campaign, it was widely believed that the stock market would benefit more from a George Bush presidency than from a Michael Dukakis presidency. Yet on January 20, 1989, the day of George Bush's inauguration as president, stock prices fell slightly. Bush's inauguration was old news. Any boost that a Bush presidency gave to the stock market happened during the election campaign as his victory became more certain. The inauguration was a well-anticipated non-event as far as the stock market was concerned.

Stock prices also go up and down because of contagious emotions— investors following the herd and being carried away by unwarranted glee or despair—Keynes' "animal spirits." Anyone who has lived through market bubbles and meltdowns knows that investors are sometimes seized *en masse* by unbridled optimism or unrestrained gloom. The thing about animal spirits is that they, too, are unpredictable. No one knows when the stock market's emotional roller coaster will start going up or when it will suddenly turn and start free-falling.

It is preposterous to think that the stock market gives away money. There is a story about two finance professors who see a hundred-dollar bill on the sidewalk. As one professor reaches for it, the other one says, "Don't bother; if it were real, someone would have picked it up by now." Finance professors are fond of saying that financial markets don't leave hundred-dollar bills on the sidewalk, meaning that if there was an easy way to make money, someone would have figured it out by now. But that doesn't deter people looking for patterns. There is just too much money to be made if a reliable pattern is found. Among the countless patterns that have been discovered (and then found to be useless) are:

- The stock market does well in years ending in 5: 2005, 2015, etc.
- The stock market does well in years ending in 8: 2008, 2018, etc.
- The stock market does well in Dragon years in the Chinese calendar.
- A New York stockbroker chose stocks by studying comic strips in a newspaper.
- A Minneapolis stockbroker let his golden retriever pick stocks.
- The Boston Snow (B.S.) indicator is based on snow in Boston on Christmas Eve.
- The Super Bowl Stock Market Predictor is based on the Super Bowl winner.

Why do people believe such nonsense? Because they are susceptible to being fooled by phantom patterns.

The Best Month to Buy Stocks

Mark Twain warned:

October: This is one of the particularly dangerous months to invest in stocks. Other dangerous months are July, January, September, April, November, May, March, June, December, August and February.

Seriously, what are the best and worst months to buy stocks? Ever-hopeful investors have ransacked past stock returns, looking for good and bad months. Investopedia advises that the "average return in October is positive historically, despite the record drops of 19.7% and 21.5% in 1929 and 1987." The January effect argues for buying in December. The Santa Claus Rally argues for buying in November. As for selling, some say: "Sell in May and go away." Others believe that August and September are the worst months for stocks.

Such advice is wishful thinking by hopeful investors searching for a way to time the market. The inconvenient truth is that there cannot be a permanent best or worst month. If December were the most profitable month for stocks, people would buy in November, temporarily making November the best month—causing people to buy in October, and so on. Any regular pattern is bound to self-destruct.

This conclusion is counter-intuitive because our lives have regular daily, weekly, and monthly cycles. The sun comes up in the morning and sets in the evening. Corn is planted in the spring and harvested in the fall. People get stronger as they grow older, then weaker as they age.

Charles Dow, the inspiration for the Dow Theory popularized by William Hamilton in his *Wall Street Journal* editorials, believed that our lives are affected by regular cycles in the economy and elsewhere. Indeed, the opening sentence of Hamilton's book, *Stock Market Barometer*, cites the sun spot theory of economic cycles. If the economy goes through regular cycles and stock prices mirror the economy, it seems plausible that stock prices should go through predictable cycles, too, and that savvy investors can profit from recognizing these cycles.

However, the fact that some businesses have seasonal patterns doesn't mean that their stocks follow such patterns. If there is a bump in toy sales before Christmas, will the price of Mattel and other toymaker stocks increase every year before Christmas? Nope. When Mattel stock trades in the summer, investors take into account the common knowledge that toy sales increase during the holiday season. They try to predict holiday sales and value Mattel stock accordingly. If their forecasts turn out to be correct, there is no reason for Mattel's stock price to change when the sales are reported. Mattel stock will surge or collapse only if sales turn out to be surprisingly strong or disappointing.

Stock prices don't go up or down because today is different from yesterday, but rather because today is not what the market expected it to be—the market is surprised. By definition, surprises cannot be predicted, so neither can short-term movements in stock prices.

Since there is no reason for a monthly pattern in surprises, there is no reason for a monthly pattern in stock prices. The patterns that are inevitably discovered by scrutinizing the past are nothing more than temporary coincidences. In the 1990s, December happened to be the best month for the stock market. In the 2000s, April was the best month and December was a nothing-burger (the sixth-best month). In the 2010s, October was the best month and April was only seventh-best. These calculations of the best months in the past are about as useful as calculating the average telephone number.

However, it doesn't stop people from making such tabulations and thinking that their calculations are meaningful. A February 2019 report from J. P. Morgan's North America Equity Research group was headlined, "Seasonality Shows Now Is the Time to Buy U. S. Gaming Stocks." The authors looked at the monthly returns for gaming stocks back to January 2000 and concluded that, "Now is the time to buy, in our view. Historically, March and April have been the best months to own U.S. gaming stocks."

Some months are bound to have higher returns than other months, just by chance. That is the nature of the unpredictable roller coaster ride we call the stock market. Identifying which month happened to have had the highest return in the past proves nothing at all.

We did a little experiment to demonstrate that monthly patterns can be phantom patterns. During the twenty-year period January 1999–December 2018, the average annualized monthly return for the S&P 500 index of stock prices was 6.65 percent, and the best month happened to be March, with a twenty-two percent average annual return. That seems like persuasive evidence that March is the best month to buy stocks.

However, suppose we were to take the 240 monthly returns over this twenty-year period and shuffle them thoroughly into twelve categories that we call pseudo-months—pseudo-January, pseudo-February, and so on. Each actual monthly return, say March 2008, would be equally likely to land in any of these pseudo-months. A return in pseudo-January is as likely to be a real June return as a real January return.

We repeated this experiment one million times. On average, the best performing pseudo-month had a twenty-three percent average return, almost exactly the same as March's real twenty-two percent return. In eighty-four percent of the simulations, there was at least one pseudo-month with an average return above twenty percent.

Remember, these are not real months. They are pseudo-months. Yes, some pseudo-months have higher average returns than others. That observation is inevitable, and useless, as is any recommendation to buy stocks based on the month of the year.

Portfolio Optimization

Historically, stocks have been very profitable investments, averaging double-digit annual returns, but stock prices are also very volatile. Figure 8.1 and Figure 8.2 show that the annual return on the S&P 500 varies a lot from year to year, much more than Treasury bonds. The S&P 500 is itself a market average and conceals an even larger turbulence in individual stocks. A stock can double or be worthless in minutes.

In the 1950s Harry Markowitz and James Tobin developed an analytical framework for taking into account both the risk and return of stock investments. Drawing on their statistical backgrounds, they proposed, as a plausible approximation, that investors base their decisions on two factors—the mean and the variance of the prospective returns. The mean is just another word for the average expected return. Since this expected value ignores uncertainty, they suggested using the statistical variance to measure risk.

Their framework has come to be known as Mean-Variance Analysis or Modern Portfolio Theory, and it has some important implications. One is that an investor's risk can be reduced by selecting a diverse portfolio of

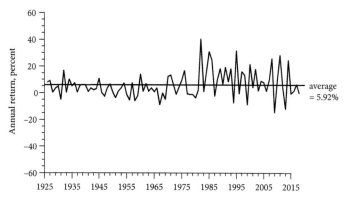

Figure 8.1 Annual returns from long-term treasury bonds since 1926.

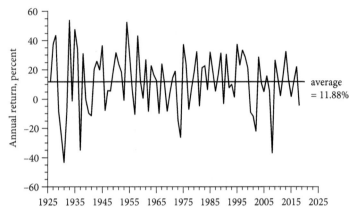

Figure 8.2 Annual returns from S&P 500 stocks since 1926.

stocks. A second implication is that risk is most effectively reduced by choosing stocks whose returns are lightly correlated, uncorrelated, or even negatively correlated with each other. A third implication is that the true measure of risk for a stock is not how much its price fluctuates, but how correlated those price fluctuations are with the fluctuations in the prices of other stocks.

A stock that performs poorly when other stocks do poorly is risky because it offers no diversification benefits. A stock that moves independently of other stocks or, even better, does well when other stocks do poorly, reduces portfolio risk. Therefore, a valid measure of risk must take into account a stock's correlation with other stocks. The Capital Asset Pricing Model (CAPM) is an offshoot of mean-variance analysis that does this.

These are valid and useful models with strong theoretical roots. The problem is that the models require data on investor expectations about the future, and the availability of historical data tempts too many to assume that the future will repeat the past—to assume that future means, variances, and correlations are equal to past means, variances, and correlations. People who rely on historical data are implicitly assuming that stocks that have done well in the past will do well in the future, and that stocks that have been relatively safe in the past will be relatively safe in the future. This is just a sophisticated form of pattern worship.

The mathematics of mean-variance analysis is intimidating and the calculations are daunting, so a whole industry has developed around turnkey portfolio management models. Financial planners and asset managers either purchase or subscribe to software services that use historical data to estimate means, variances, and correlations and recommend portfolios based on these historical data.

Historical returns can be a misleading guide to the future and lead to unbalanced portfolios that are heavily invested in a small number of stocks that happened to have done well in the past, the opposite of the diversification recommended by mean-variance analysis.

For example, in 2019, Gary noticed a website that offered to train people to become financial data scientists. The website was polished and included an interactive demonstration of a portfolio optimization program that had been presented at an R/Finance Conference for people who use the R programming language to create financial algorithms. The site boasted that "We can use data-driven analysis to optimize the allocation of investments among the basket of stocks."

The site mentioned impressive buzzwords like Modern Portfolio Theory and the Capital Asset Pricing Model, and claimed that "the investment allocation decision becomes automated using modern portfolio theory…[The program] helps an Asset Manager *make better investment decisions that will consistently improve financial performance and thus retain clients*" (bold and italics in original).

Gary tried out this portfolio optimization program. The only inputs that are required are the names of three stocks. What could be simpler than that? Asset managers do not need to make predictions about future performance because the program assumes that future performance will be the same as past performance. If a fly-by-night stock happened to do really well in the past, the optimization program would recommend investing heavily.

Gary tested the program with three rock-solid companies (no fly-by-nights): Google, IBM, and Microsoft. Looking at the performance of these three stocks during the five-year period, 2009 through 2013, the optimizer program recommended a portfolio of sixty-three percent IBM, twenty-three percent Google, and fourteen percent Microsoft. The program reported that this portfolio beat the S&P 500 by four percentage points per year, with less volatility.

It was tempting to think that this portfolio would "make better investment decisions that will consistently improve financial performance and thus retain clients" because it had been selected by a fancy algorithm. However, this portfolio was not based on anyone's analysis of the future. It was just a report that this portfolio had beaten the S&P 500 by four percentage points per year during the years 2009 through 2013. This is known as *backtesting*, and it is hazardous to your wealth.

It is easy, after the fact, to identify strategies that would have done well in the past, but past performance has very little to do with future performance in the stock market. IBM got the heaviest weight and Microsoft the smallest because IBM had performed the best, and Microsoft the worst, during the years 2009 through 2013. It is very much like noticing a pattern in coin flips and assuming that the pattern will continue.

How did this IBM-heavy portfolio do over the next five years, 2014 through 2018? It *underperformed* the S&P 500 by two percentage points per year because, as Figure 8.3 shows, Microsoft tripled in value, while IBM lost twenty-eight percent of its value. Based on this new five-year performance, the optimization program flipped its recommendation from sixty-three percent IBM and fourteen percent Microsoft to seventy-six percent Microsoft and only four percent IBM. Alas, that is still a report about the past, not a reliable prediction about the future.

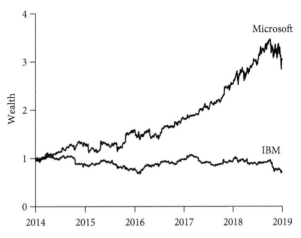

Figure 8.3 IBM was not the better investment.

It is easy to identify stocks that have done well, but difficult to pick stocks that will do well. Assuming that past stock performance is a reliable guide to future performance is being fooled by phantom patterns.

A System of the Month

Gary recently received an e-mail solicitation offering access to hundreds of automated trading algorithms: "Like an Algo App store." The company claimed that a day-trading "System of the Month" would have made an $82,000 profit on a $2,600 investment over the previous three-and-a-half years. The cost was $70 a month, which works out to $840 a year (forty-two percent of the initial investment!) plus a $7.50 commission on each trade (and day-trading systems have lots of trades). The system must have been incredibly successful to overcome its steep costs.

The system was said to be fully automated, hands-free, unemotional, and backtested.

Well, there's the gimmick. It was based on backtesting. Backtested systems do spectacularly well because it is always possible to find a system that would have done well in the past, if it had been known ahead of time. The fine print admits as much:

NO REPRESENTATION IS BEING MADE THAT ANY ACCOUNT WILL OR IS LIKELY TO ACHIEVE PROFITS OR LOSSES SIMILAR TO THOSE SHOWN. IN FACT, THERE ARE FREQUENTLY SHARP DIFFERENCES BETWEEN HYPOTHETICAL PERFORMANCE RESULTS AND THE ACTUAL RESULTS SUBSEQUENTLY ACHIEVED BY ANY PARTICULAR TRADING PROGRAM.

ONE OF THE LIMITATIONS OF HYPOTHETICAL PERFORMANCE RESULTS IS THAT THEY ARE GENERALLY PREPARED WITH THE BENEFIT OF HINDSIGHT.

Duh!

Rolling Dice for Dollars

Some companies sell computerized systems for beating the market. Others recruit investors who want their money managed by computerized systems. In both cases, the results are usually disappointing.

An exchange-traded fund (ETF) is a bundle of securities, like a mutual fund, that is traded on the New York Stock Exchange and other exchanges like an ordinary stock. There is now nearly $5 trillion invested in 5,000 ETFs.

One attractive feature is that, unlike shares in ordinary mutual funds, which can only be purchased or sold after the markets close each day, ETFs can be traded continuously while markets are open. This evidently appeals to investors who think that they can jump in and out of the market nimbly, buying before prices go up and selling before prices go down. Unfortunately, day trading takes a lot of time and the outcomes are often disheartening. Perhaps computers can do better?

Artificial intelligence (AI) has become a go-to word. In 2017 the Association of National Advertisers chose "AI" as the Marketing Word of the Year. In October of that year, a company with the cool name EquBot launched AIEQ, which claimed to be the first ETF run by AI, and not just any AI, but AI using Watson, IBM's powerful computer system that defeated the best human players at Jeopardy.

Get it? AI stands for artificial intelligence and EQ stands for equity (stock). Put them together and you have AIEQ, the artificial intelligence stock picker. EquBot boasts that AIEQ is "the ground-breaking application of three forms of AI:" genetic algorithms, fuzzy logic, and adaptive tuning. Investors may not know what any of these words mean, but that is part of the allure. If someone says something we don't understand, it is natural to think that they are smarter than us. Sometimes, however, mysterious words and cryptic phrases are meant to impress, not inform.

Chida Khatua, CEO and co-founder of EquBot, used ordinary English, but was still vague about the details: "EquBot AI Technology with Watson has the ability to mimic an army of equity research analysts working around the clock, 365 days a year, while removing human error and bias from the process." We remember Warren Buffett's advice to never invest in something you don't understand. We also acknowledge that it would be nice to remove human error and bias (Is anyone in favor of error and bias?), but computers looking for patterns have errors and biases, too.

Computer algorithms for screening job applicants, pricing car insurance, approving loan applications, and determining prison sentences have all had significant errors and biases, due not to programmer biases, but rather to the nature of patterns. An Amazon algorithm for evaluating job applicants discriminated against women who had gone to women's colleges or belonged to women's organizations because there were few women in the algorithm's database of current employees. An Admiral

Insurance algorithm for setting car insurance rates was based on an applicant's Facebook posts. One example the company cited was whether a person liked Michael Jordan or Leonard Cohen—which humans would recognize as ripe with errors and biases.

Admiral said that its algorithm:

is constantly changing with new evidence that we obtain from the data. As such our calculations reflect how drivers generally behave on social media, and how predictive that is, as opposed to fixed assumptions about what a safe driver may look like.

This claim was intended to show that their algorithm is flexible and innovative. What it actually reveals is that their algorithm finds historical patterns, not useful predictors. The algorithm changes constantly because it has no logical basis and is continuously buffeted by short-lived correlations.

Algorithms based on patterns are inherently prone to discover meaningless coincidences that human wisdom and common sense would recognize as such. Taking humans completely out of the process and hiding the mindless pattern discovery inside an inscrutable AI black box is not likely to end well.

Figure 8.4 shows how AIEQ worked out. Despite the promises of genetic algorithms, fuzzy logic, and adaptive tuning, AIEQ seems to be a "closet

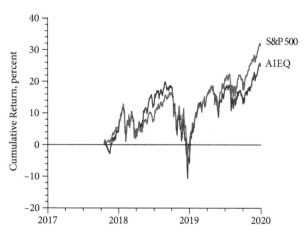

Figure 8.4 Underwhelming performance.

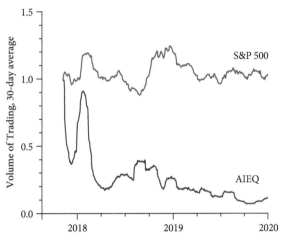

Figure 8.5 Overwhelming disinterest.

indexer," tracking the S&P 500, while underperforming it. From inception through November 1, 2019, AIEQ had a cumulative return of twenty-five percent, compared to thiry-two percent for the S&P 500. Meanwhile, the company collected a 0.77 percent annual management fee for this distinctly mediocre performance.

Figure 8.5 compares the volume of trading in AIEQ to the volume of trading in the S&P 500, both scaled to equal 1 when AEIQ was launched. Once the disappointing results became apparent, customers lost interest.

In investing and elsewhere, an "AI" label is often more effective for marketing than for performance.

Where's the Beef?

Hopeful investors are not the only ones who have fallen in love with AI. In 2017, the year "AI" was anointed as the Marketing Word of the Year and AIEQ was launched, a startup named One Concern raised $20 million to "future proof the world."

One Concern had been inspired by the experience of a Stanford graduate student (Ahmad Wani), who had been visiting his family in Kashmir when a flood hit. His family had to wait seven days to be rescued from their flooded third-story apartment and they saw neighbors killed by collapsing homes.

When Wani returned to Stanford, he worked with two other students, Nicole Hu and Tim Frank, on a class project to predict how buildings would be affected by earthquakes, which are more common than floods in Silicon Valley. They turned that project into a company, One Concern, and reported that their AI algorithm can estimate, in less than fifteen minutes after an earthquake hits, the block-by-block damage with eighty-five percent accuracy. In a *Time* magazine interview, Wani said that, "Our mission is to save lives. How do we make the best decisions to save the most lives?"

This inspiring story was widely reported and One Concern soon had its first $20 million in funding (it has received much more since then) and several customers, including San Francisco and Los Angeles. It has since launched Flood Concern (evidently there is more than one concern), a companion product for floods.

When it received an initial $20 million in funding, Wani wrote that:

Our AI-based technology will assign a unique, verified 'digital fingerprint' to every natural or manmade element, from the smallest rock to complete structures to mega cities and eventually, the entire planet. One Concern will provide insights across the entire time horizon—whether it's days before a flood, minutes after an earthquake, or forward-looking policy and planning.

A skeptic might be forgiven for thinking that this vision is more sizzle than steak, a skepticism buttressed by this promise on their slick website:

Our long-term vision is for planetary-scale resilience where everyone lives in a safe, sustainable and equitable world.

Ms. Hu has stated that, "We really think we can enable a disaster-free future."

A successful algorithm might help guide rescuers to the places they are needed, but how will improved rescue operations eliminate disasters and ensure that "everyone lives in a safe, sustainable and equitable world"? Floods and earthquakes will still kill, and what is the relevance of the buzzwords "sustainable and equitable world"?

The only measurable claim is that their algorithm can predict block-by-block damage with eighty-five percent accuracy. The number "85" is well-chosen. It is not so low as to be disappointing, nor so high as to be unbelievable. It is more precise than "80" or "90", suggesting that it is based on a scientific study.

What does eighty-five percent accuracy mean? If it refers to the amount of block-by-block damage, eighty-five percent accuracy is meaningless. If it refers to predictions of which block is the most damaged, how do we aggregate injuries, lives, and buildings into a measure of damage?

Perhaps it means that, of those buildings that the algorithm identifies as damaged, eighty-five percent turn out to be damaged. For example, suppose that eighty-five percent of the buildings in an area are damaged and the algorithm picks 100 buildings at random, of which eighty-five were damaged. That could be interpreted as eighty-five percent accuracy, even though the algorithm is useless. Or suppose that 100 out of 1,000 buildings in an area are damaged, and the algorithm picks 850 buildings at random, of which eighty-five are actually damaged. That, too, could be interpreted as eighty-five percent accuracy even though the algorithm is useless.

The algorithm should be compared to predictions made without the algorithm—a control group. Does it do better than experienced local rescue workers who are familiar with the area, or helicopters flying over the area? When Wani waited seven days to be rescued in Kashmir, perhaps the problem was not that no one knew his neighborhood had been flooded, but that there were not enough rescue resources.

Maybe One Concern is all sizzle and no steak.

In August 2019, *The New York Times* published an article, "This High-Tech Solution to Disaster Response May Be Too Good to Be True," written by Sheri Fink, who has won one Pulitzer Prize for Investigative Reporting and shared another for International Reporting.

San Francisco had ended its contract with One Concern and Seattle had doubts about the program's cost and reliability. In one test simulation, the program ignored a large Costco warehouse, because the program relies mainly on residential data. The program also had errors in its strength assessments of buildings and counted every apartment in a building as a separate structure. When One Concern revised the program to include the missing Costco, the University of Washington mysteriously vanished. With each iteration, the damage predictions changed dramatically.

One Concern persuaded an insurance company to pay $250,000 for Seattle to use the algorithm. In return, the insurance company is able to use the model's predictions to assist the "design of our insurance products as well as the pricing." That sure sounds like the company is looking for reasons to raise rates.

One former employee told Fink that the eighty-five percent accuracy number was misleading and that, "One of the major harms is the potential to divert attention from people who actually need assistance." Another said he was fired for criticizing the company's dishonest attitude of "fail fast and try something new" because, with disaster response, "If you fail fast, people die."

Mr. Wani's response was revealing: "We are in no way ever telling these first responders that we are replacing your decision-making judgment or capability." Isn't that exactly what they were advertising?

Others pointed out that no tests of the algorithm have been published in peer-reviewed journals and that the AI label is misleading. Genuine AI programs train on vast quantities of data, and there are relatively little data on earthquakes of comparable magnitudes in comparable locations. If the algorithms were extrapolating data from earthquakes of certain magnitudes to earthquakes of other magnitudes, from earthquakes in certain geographic locations to other locations, and from certain types of buildings to other buildings, then they were being fooled by phantom patterns. How relevant is building damage data from a shallow 7.3 earthquake in Indonesia for predicting the damage from a deep 6.5 earthquake in San Francisco?

Mr. Wani eventually revised the eighty-five percent accurate number to seventy-eight percent, but was vague about what it means: "You know, we don't even call it 'accuracy'; we call it a 'key performance indicator.' If you have to send first responders to respond after the disaster for, let's say, carrying out urban search and rescue, you'd be at least seventy-eight percent or higher, or at least more than seventy-eight percent accurate for doing that." The number is different, but still meaningless.

A One Concern simulation predicted that the earth under one California highway would liquify in a major earthquake, but that information is readily available from the state. Similarly, the simulations rely on damage predictions made by another company using free government data, with the CEO stating bluntly, "It's not AI."

One Concern has increasingly been working with insurance companies and credit-rating agencies that might use One Concern's predictions to raise individual insurance rates and downgrade city bonds. Instead of saving lives, their focus seems to have shifted to justifying higher insurance premiums and loan rates. Based on an interview with a former employee, Fink wrote that:

The shift in the financial model "felt very deceitful" and left many employees feeling disillusioned, said Karine Ponce, a former executive assistant and office manager.

An article in *Fast Company*, following up on an earlier positive article, concluded that:

As the *Times* investigation shows, the startup's early success—with $55 million in venture capital funding, advisors such as a retired general, the former CIA director David Petraeus, and team members like the former FEMA head Craig Fugate—was built on misleading claims and sleek design…With faulty technology that is reportedly not as accurate as the company says, One Concern could be putting people's lives at risk.

Shopping for Dollars

Tesco is a British-based company that grew from a small supermarket chain into one of the world's largest retailers.

The transformation took off in 1994 when Tesco initiated a Clubcard loyalty program. Shoppers with Clubcards earn points that can be redeemed for merchandise at Tesco and partner restaurants, hotels, and other locations. The traditional corporate appeal of loyalty programs is to lock in customers. People who are eligible for a free coffee after buying nine coffees from Julie's Java are more likely to buy coffee at Julie's.

Tesco flipped the script. The real value of the Clubcard for Tesco is the detailed information it provides about its shoppers—the same way that Target was able to use its customers' shopping habits to identify pregnancies and predict birth dates in order to "target" them with special offers. Tesco contracted with the data mining firm dunnhumby (the quirky British name with no upper-case letters coming from the wife and husband, Edwina Dunn and Clive Humby, who founded the firm in their kitchen in 1989). When dunnhumby presented its initial analysis to the Tesco board in 1994, the board chair famously said, "What scares me about this is that you know more about my customers after three months than I know after 30 years."

Tesco later bought dunnhumby and went all in. Today, half of all British households have a Clubcard. The Tesco data analysts use Clubcard data together with social media data to predict which deals will appeal most to individual customers. One customer might be attracted to discounts on soap, but care more about quality for tomatoes; another customer may feel the exact opposite. So, the first customer is sent discount coupons for soap, while the second customer is sent coupons for tomatoes.

Tesco's profits quintupled between 1994 and 2007 and its stock price increased by a factor of ten. In November 2007, Tesco began opening Fresh & Easy stores in the United States, confident that their data prowess would ensure their success. CEO Terry Leahy boasted that, "We can research and design the perfect store for the American consumer in the 21st century. We did all our research, and we're good at research."

Tesco's researchers decided that American consumers wanted one-stop shopping where they could buy organic products at fair prices in an environmentally friendly store. So, they built clean, well-lit stores with organic food and solar roofing.

By the end of 2012, there were more than 200 Fresh & Easy stories in the western United States. It did not end well. In April 2013, Tesco announced that it was giving up on the U.S. market. The total cost of this fiasco was $3 billion.

One data problem was that Americans said that they like one-stop shopping but, as an American politician once advised, "Watch what we do, not what we say." Many Americans have a love affair with their cars and enjoy driving around from one store to another. Americans may like clean, well-lit stores, but they also enjoy shopping at quirky places like Jungle Jim's, Kalustyan's, Rouses, Trader Joe's, and Wegmans. Whatever the reason, Tesco's data gurus badly misread the American market.

Figure 8.6 and Figure 8.7 show the subsequent cratering of Tesco's profits and stock price. In addition to its botched invasion of the U.S. market, Tesco had some accounting issues and fierce competition from aggressively low-cost firms like Aldi, a German supermarket chain, but a 2014 *Harvard Business Review* article argued that data missteps were an important part of the problem: "In less than a decade, the driver and determinant of Tesco's success has devolved into an analytic albatross."

One problem with Clubcard discounts, a problem that is very familiar to pizza stores, is that once customers get accustomed to coupons, they won't buy without a coupon. For example, a household that tends to have pizza once a week, might buy pizza on a second day of the week if they see an irresistible discount—which boosts the pizza store's revenue that week. It might happen again a second week, and then a third week, but now the household is using discounts twice a week, which eats into profits. Then

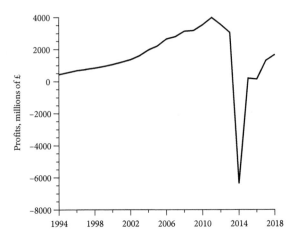

Figure 8.6 Profits falling off a cliff.

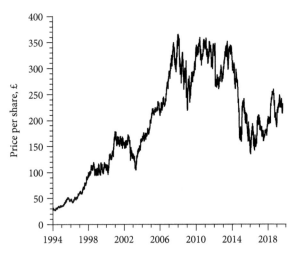

Figure 8.7 The Tesco stock price roller coaster.

the household comes to consider discount coupons the new normal and goes back to having pizza once a week, but only if there is a coupon.

Some households think that coupons are gimmicky and a bother. They would rather shop where prices are always low instead of having to collect and redeem coupons. Other customers game the system once they figure it out. It's like digital haggling. The store sends a discount coupon for tomatoes and the customer doesn't bite—perhaps because they forget to use it. The next thing you know, the store sends a bigger discount, trying to find the customer's price point. The customer is now wise to the game, and waits for a third or fourth coupon before making a move.

We've seen this happen with Amazon. A customer looks at a product for several minutes, checking the specifications and reading the reviews, and then doesn't buy. The next time, the customer looks at the product, the price is mysteriously lower. The reverse happens too. A customer orders a bag of coffee bean several months in a row and then notices that the price has mysteriously increased. Amazon apparently hopes they won't notice. The customer then asks a friend or relative to order the beans for them, and they are quoted the original price.

Another problem with targeted coupons is cannibalization. A soap coupon for one customer in one Tesco store might boost sales in that store at the expense of sales in another Tesco store.

A more general problem is that Tesco only has data for people with Clubcards who shop at Tesco. Tesco doesn't know much about Tesco shoppers who do not have cards, and it knows nothing about the purchases that people (Clubcard or not) make in other stores. For example, Tesco might notice Clubcard holders paying premium prices for heirloom tomatoes in some of its stores, but it doesn't notice that most shoppers are buying less expensive tomatoes elsewhere. If new households come into an area and shop at non-Tesco stores, the Tesco data analysts will have no idea what they are buying or why.

As with all pattern-seeking algorithms, Tesco's number crunching is also vulnerable to coincidences; for example, an algorithm may notice that people who bought tomatoes on Thursday in one store happened to have bought toothpaste too, and conclude that this coincidence means something.

Another data problem is that although micro-targeting is alluring and potentially profitable, it is possible to make the proverbial mistake of missing the forest for the trees. A micro-targeting algorithm that notices that individual customers are buying less sweetened cereal might use coupons to try to lure these customers to buy more, while not noticing that the store's clientele are less interested in sweetened cereal because they are getting older. The real challenge is not to sell more sweetened cereal to older customers, but to find ways to attract younger shoppers. Similarly, individual customers buying less beef may be a sign that many people are cutting back on beef and would prefer not more beef coupons, but more beef alternatives.

Finally, as we have come to realize how closely our behavior is being monitored, many have come to resent it. Facebook's Mark Zuckerberg once called people who trusted him with their data "dumb f**ks." Some of these dummies are waking up and not taking it anymore. A 2019 study found that the number of U.S. Facebook users dropped by 15 million between 2017 and 2019, though many of these fleers fled to Facebook-owned Instagram.

Nordstrom stopped using wi-fi sensors to track customers in their stores the day after a CBS affiliate reported the snooping. Urban Outfitters has been hit with a class-action lawsuit for falsely telling shoppers paying by credit card that they have to tell the store their ZIP code, which can then be used to determine their home addresses. A 2013 study found that one-third of those surveyed reported that they had stopped using a website or buying from a company because of privacy issues. Some people have even gone back to paying cash and not buying anything over the Internet.

In 2018, Apple's Tim Cook warned that:

Our own information—from the everyday to the deeply personal—is being weaponized against us with military efficiency…Every day billions of dollars change hands and countless decisions are made on the basis of our likes and dislikes, our friends and families, our relationships and conversations, our wishes and fears, our hopes and dreams…We shouldn't sugarcoat the consequences. This is surveillance…For artificial intelligence to be truly smart it must respect human values—including privacy. If we get this wrong, the dangers are profound.

A British newspaper wrote of Tesco's problems:

judging by correspondence from Telegraph readers and disillusioned shoppers, one of the reasons that consumers are turning to [discounters] Aldi and Lidl is that they feel they are simple and free of gimmicks. Shoppers are questioning whether loyalty cards, such as Clubcard, are more helpful to the supermarket than they are to the shopper.

Customer disenchantment with Tesco's Clubcard may also reflect a growing distrust of data collectors and a growing wish for privacy. Some Target customers were upset about Target's pregnancy predictor algorithm (so

Target started adding random products like lawnmowers in their ads to hide how much they knew). Tesco's customers may reasonably not want the world to know how much alcohol, laxatives, condoms, and Viagra they buy.

In 2015, Tesco put dunnhumby up for sale, but did not get any attractive offers.

How to Avoid Being Misled by Phantom Patterns

Data are undeniably useful for answering many interesting and important questions, but data alone are not enough. Data without theory has been the source of a large (and growing) number of data miscues, missteps, and mishaps.

We should resist the temptation to believe that data can answer all questions, and that more data means more reliable answers. Data can have errors and omissions or be irrelevant, which is why being duped by data is a corollary of being fooled by phantom patterns. In addition, patterns discovered in the past will vanish in the future unless there is an underlying reason for the pattern.

Backtesting models in the stock market is particularly pernicious because it is so easy to find coincidental patterns that turn out to be expensive mistakes. This endemic problem has now spread far and wide because there are so much data that can be used by academic, business, and government researchers to discover phantom patterns.

CHAPTER 9

Seeing Things for What They Are

I n January 1994, just as the World Wide Web (WWW) was starting to get traction, two Stanford graduate students, Jerry Yang and David Filo, started a website named "Jerry and David's guide to the World Wide Web." Their guide was a list of what they considered to be interesting web pages. A year later, they incorporated the company with the sexy name Yahoo!, emboldened with an exclamation point. How many company names contain exclamation points? They now had a catalog of 10,000 sites, and 100,000 Yahoo users a day, fueled by the fact that the popular Netscape browser had a Directory button that sent people to Yahoo.com.

As the web took off, Yahoo hired hundreds of people to search the web for sites to add to its exponentially growing directory. They added graphics, news stories, and advertisements. By 1996, Yahoo had more than ten million visitors a day. Yahoo thought of itself as a media company, sort of like *Fortune* magazine, where people came to be informed and entertained, and advertisers paid for a chance to catch readers' wandering eyes. Yahoo hired its own writers to create unique content. As Yahoo added more content, such as sports and finance pages, it attracted targeted advertising—directed at people who are interested in sports or finance. It started Yahoo mail, Yahoo shopping, and other bolt-ons.

Then came Google in 1998 with its revolutionary search algorithm. Yahoo couldn't manually keep up with the growth of the web, so it used Google's search engine for four years while it developed its own algorithm. Meanwhile, Google established itself as the premiere search site and, today, still maintains a ninety percent share of the search market.

Yahoo made some colossal blunders in 1999 by paying $3.6 billion for GeoCities and $5.7 billion for Broadcast.com. (In contrast, Google paid $50 million for Android in 2005, $1.65 billion for YouTube in 2006, and $3.1 billion for DoubleClick in 2008.) Still, people went to Yahoo's pages because it had great content and because of user inertia. Yahoo stock soared to a split-adjusted peak of $118.75 in January 2000. Unlike most dot-coms, Yahoo was profitable. Still, it only earned five cents a share. Stocks are generally considered dangerously expensive when the price/earnings (P/E) ratio goes above, say, 50. Yahoo's P/E was a mind-boggling 2,375. By one estimate, to justify its market valuation, Yahoo would have to be as profitable as Wal-Mart in 2000, twice as profitable in 2001, three times as profitable in 2002, and so on, forever.

Yahoo was deliriously overvalued. The stock market soon came to its collective senses and Yahoo fell off a proverbial cliff. Yahoo stock plummeted ninety percent over the next twelve months.

The Yahoo stock crash was, no doubt, an over-reaction (Wall Street is famous for that), and its stock climbed back up over the next several years. During this time, Yahoo went through a merry-go-round of CEOs. None of them seemed to make a difference, but they were paid outrageous amounts.

Charisma is rarely the solution to a company's problems. As Warren Buffett put it, "When a manager with a great reputation meets a company with a bad reputation, it is the company whose reputation stays intact."

Disappointment is likely to be especially acute when an outsider is brought in as CEO. First, an outsider doesn't know the company's culture and the strengths and weaknesses of its employees. Insiders know that Jack is a complainer, that Jill is a boaster, and that John is a slacker. Outsiders don't know Jack, Jill, or John. Second, the board making the hiring decision doesn't know an outsider as well as it knows its own insiders. The less information there is, the wider the gap between perception and reality.

Yahoo learned these lessons the hard way, hiring five CEOs (four of them outsiders) in five years in a futile attempt to save a slowly sinking ship.

With Yahoo's stock down to $10.47 a share, Yahoo's original CEO, Tim Koogle, was fired on March 7, 2001, and replaced by Terry Semel on April 17, 2001. Semel had worked at Warner Brothers for twenty-five years. He had been both chairman and co-chief executive officer, and was known for his Hollywood deal-making.

Unfortunately, his Hollywood background didn't play well in Silicon Valley. Semel wanted to build a Yahoo entertainment unit, and run Yahoo as if it were a movie studio, like the Warner Brothers studio he was comfortable with. He botched several deals, including walking away from opportunities to buy Google and Facebook. Google wanted $3 billion; Semel wouldn't go above $1 billion. Semel had a deal to buy Facebook for $1 billion, but he tried to cut the price to $800 million at the last minute. Semel also reportedly had an agreement to buy DoubleClick for $2.2 billion which fell through at the last minute. On the other side of the table, Microsoft offered to buy Yahoo, but was rebuffed.

Semel shows up on several lists of the worst tech CEOs ever, yet he was paid nearly $500 million for his six years as CEO. In 2007, Yahoo's board of directors gave Semel a vote of confidence, then fired him a week later.

Yahoo then promoted Jerry Yang, one of the company's co-founders, and, like Google co-founders Sergey Brin and Larry Page, Yang took a token salary of $1 a year (his Yahoo stock was worth billions). However, most of his tenure probably wasn't worth a dollar. Yang lasted a little over a year, and it was not a good year. Yahoo's earnings dropped twelve percent and its stock price fell sixty percent. He also rebuffed Microsoft's offer to buy Yahoo for $44.6 billion (sixty-two percent over its market price). Besides co-founding the company, Yang's biggest accomplishment occurred before he became CEO, when he spearheaded the 2005 purchase of a forty percent stake in an Internet start-up named Alibaba for $1 billion. Alibaba has become the Amazon, eBay, and Google of China. Since Alibaba shares were not yet publicly traded, some investors bought Yahoo stock just to get an investment in Alibaba.

After Yang stepped down, Carol Bartz became CEO. She had been CEO of Autodesk, which makes design software, and announced that she was coming to Yahoo to "kick some butt." She laid off hundreds of workers during her twenty-month term, but Yahoo stock only went up seven percent during a period when the NASDAQ went up more than sixty percent. After she was paid $47.2 million in 2010, the proxy-voting firm Glass Lewis called her the most-overpaid CEO in the country. She was fired over the phone in September 2011.

Next up was Scott Thompson, who had been president of PayPal. He laid off another 2,000 Yahoo employees in four months before he was let go amidst allegations that he falsely claimed to have a degree in computer science. He was paid more than $7 million for 130 days as CEO.

Ross Levinsohn, a Yahoo executive vice president, agreed to serve as interim CEO and did so for three months. He left with essentially a $5 million severance package after Marissa Mayer was hired as CEO on July 16, 2012.

Mayer was only thirty-seven years old, but she was smart and energetic and had an incredible resume. She had been a dancer, debater, volunteer, and teacher at Stanford, while earning BS and MS degrees. She had several artificial intelligence patents and was employee number 20 at Google, where she quickly rose to a vice-president position. In 2012, Yahoo hired Mayer as president and CEO to reverse Yahoo's long downward spiral (caused, in part, by the rapid ascent of Google).

Mayer set out to restore Yahoo's glory, to make it a player like Amazon, Apple, Facebook, and Google. It wouldn't be easy. Yahoo's search engine had only a small share of the market. Yahoo's news content was valuable, but users couldn't access it easily on their smartphones. Yahoo Mail carried billions of e-mails every day, but users couldn't read them easily on their smartphones.

Mayer's plan was (a) develop Yahoo mobile apps for news, e-mail, and photo-sharing; (b) create must-read original content to draw readers into the Yahoo universe; (c) improve Yahoo's search algorithm; and (d) make Yahoo again a fun and exciting place to work, a place that would attract and nurture good talent.

She told Yahoo employees that, "Our purpose is to inspire and delight our users, to build beautiful services, things that people love to use and enjoy using every day, and that's our opportunity."

In her first two years, Mayer eliminated old products, created new ones, and bought forty-one start-ups—in part, for an infusion of new talent and creativity. She argued that "Acquisitions have not been a choice for Yahoo in my view but, rather, a necessity." Yahoo tried to offer new Internet services that would lure users so that advertisers would follow, but it couldn't compete effectively against Craigslist, eBay, Facebook, and Google. Yahoo's main distinguishing service was premium content—which requires expensive human labor.

She hired Henrique de Castro away from Google in November 2012 to be Yahoo's chief operating officer (COO) by making him the highest paid COO in the country. De Castro was said to have been abrasive at Google and abrasive at Yahoo. Advertising revenue fell every quarter after his hire and he was fired in January 2014 with a $60 million severance package. His total compensation for fifteen disappointing months was $109 million.

None of Mayer's goals were met. The apps were lackluster, the digital magazines were unpopular, and the search engine lagged even further behind Google.

Gary used to be a Yahoo devotee because of the great content. Unlike the initial Google site, which was just a search engine, Yahoo had links to important news stories from *The New York Times*, *Wall Street Journal*, BBC, and other reputable sources. They also had well organized sections on sports and finance, with well-organized data and interesting stories, some written by Yahoo's own staff. Gary could look up any stock he was interested in and find a treasure trove of interesting information.

Then someone decided that Yahoo could increase its advertising revenue by planting click bait, which are ads disguised as news stories, like these examples: "What This Child Actress Looks Like Now Will Drop Your Jaw!," "The Recluse Who Became the Greatest Stock Picker," and "Horrifying Woodstock Photos That Were Classified."

A user who clicks on the Woodstock link is taken to a slide show labeled "Forgotten Woodstock: Never Seen Before Images of the Greatest Rock Concert of all Time!" This slide show is surrounded by advertisements and more "stories" with titles like "Incredibly Awkward Family Photos," "Unbelievable Things That Really Happened," and "She Had No Idea Why the Other Politicians Stared" (accompanied by a photo of a large-breasted woman). Clicking on any of these stories brings up yet another slide show that is surrounded by more advertisements and "stories." It is difficult to navigate through any of the slide shows without hitting a "Next" button or arrow that takes the user to an advertisement. The slide shows are utterly boring. (Gary only made it through four of the Woodstock slides.)

The legitimate stories that remain on Yahoo's main page are mostly celebrity fluff: two movie stars break up; two celebrities are seen together; a school teacher has an affair with a student; there was a baby mix-up at the hospital.

At the same time that Yahoo was evolving into a combination of the National Enquirer and the Yellow Pages, Google added a news page with real news stories from CNN, *The New York Times*, *Wall Street Journal*, and *Chicago Tribune* about things that are more important than celebrity gossip. There might be stories about flooding in Paris, thousands in Hong Kong commemorating the Tiananmen Square massacre, and rumors that Facebook is monitoring smartphone conversations.

Google News is now Gary's go-to web page.

In 2016 Mayer announced that Yahoo would lay off fifteen percent of its 11,000 employees and shut down half of its digital publications: Yahoo Autos, Yahoo Food, Yahoo Health, Yahoo Makers, Yahoo Parenting, Yahoo Real Estate, and Yahoo Travel. Yahoo employees were reportedly demoralized by the shift from quality content to low-brow cheap thrills on "a crap home page." A common complaint from the writers was said to be "You are competing against Kim Kardashian's ass."

Yahoo's stock price continued to fall and key employees fled.

Yahoo got a $9.4 billion cash infusion in 2014 by selling twenty-seven percent of its Alibaba shares. In 2016 Yahoo's remaining 383 million Alibaba shares (fifteen percent of Alibaba) were reportedly the only thing of value that Yahoo had left. Outside analysts estimated that the market value of Yahoo was less than the value of its Alibaba holdings—Yahoo's business had a negative market value! Investors told Yahoo to split the company in two, one part the valuable Alibaba shares, the other part the essentially worthless Yahoo business. Yahoo's collapse was seemingly complete, from a $128 billion-dollar company to nothing. In the summer of 2016, Yahoo sold its core business, excluding Alibaba, to Verizon for $4.8 billion, which is more than zero, but not a lot more.

For presiding over this sinking ship, it has been estimated that Mayer was paid more than $200 million, including a $57 million severance package.

Figure 9.1 shows the history of Yahoo's stock price, adjusted for stock splits, along with the timelines of the five CEOs who were brought in to save Yahoo after its stock price crashed. (Ross Levinsohn's three-month interim position is omitted).

Five highly touted and richly rewarded CEOs collectively couldn't do much for Yahoo beyond the fortuitous purchase of Alibaba stock, which was spearheaded by Jerry Yang during a time when he was not CEO. For this, they were paid almost a billion dollars. In every case, Yahoo paid a small fortune to bring in a superstar CEO who would restore Yahoo's glory. In every case, the gap between hope and reality turned out to be enormous.

This sad story is an example of the fact that phantom patterns don't have to be numerical. Every CEO who walked through Yahoo's revolving door had been successful in the past. Yahoo's board assumed that a past pattern of success was a reliable predictor of future successes. However, a

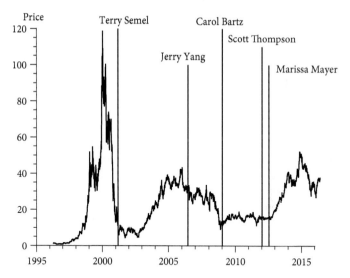

Figure 9.1 Yahoo!'s rise and fall.

CEO is seldom the reason for a company's success. It is the thousands or tens of thousands of employees coming to work every day who make a company great. There can also be a large dollop of luck. Some companies prosper because a product turns out to be far more successful than anyone could reasonably have anticipated. Who knew that frisbees, hula hoops, and Rubik's cubes would be so popular? Who knew how successful Post-it Notes and Toyota Corollas would be? Not many. Certainly not the CEOs.

Sometimes, a company will do well in spite of its CEO. No need to name names. The point is that the simple pattern—Chris Charisma becomes CEO, company does well—is not a reliable predictor of how well the next company that hires Charisma will do.

The Peter Principle

It is not just CEOs. Employees are usually promoted based on their outstanding performance in their current jobs. The jobs they are promoted to, however, may require very different skills; for example, being promoted from engineer to supervisor or from salesperson to sales manager. This is why people who are promoted based on their current performance, instead of the abilities they need to succeed in their new job, are often disappointing.

Promoted employees who do turn out to be successful in their new jobs are likely to be promoted to an even higher level in the hierarchy. And so it goes, until they reach a position where they are not successful and not promoted any further up the chain.

The promotion of people who are good at what they did until they are no longer good at what they now do is the Peter Principle, coined by Laurence J. Peter: "managers rise to the level of their incompetence."

The Peter Principle implies that most managers are ineffective, and those employees who are effective are only effective until they get promoted to jobs where they are ineffective. The Peter Principle is a cynical, but all-too-common, reality.

Google

A co-worker at Jay's last company was asked how they could be successful as an Internet company when Google is the big fish in the sea. He answered: "Google's not the big fish in the sea, they *are* the sea." So, how did Google become the sea?

Larry Page graduated from the University of Michigan; Sergey Brin was born in Moscow and graduated from the University of Maryland. Both dropped out of Stanford's Ph.D. program in computer science to start Google. Their initial insight was that a web page's importance could be gauged by the number of links to the page. Instead of paying humans to guess what pages a searcher might be interested in, they used the collective votes of Internet users to determine the most relevant and useful pages. To collect these data, they created a powerful web crawler that could roam the web counting links. This search algorithm, called PageRank, has morphed into other, more sophisticated algorithms, allowing Google to dominate the search market.

Google's bots (called *spiders* or *crawlers*) are constantly roaming the web, collecting data for their matching algorithms. In 2019, it was estimated that Google handled 63,000 searches per second, or 5.5 billion searches per day, which it matches to data from sixty billion web pages. This enormous database of searchers and search results allows Google to continually improve its algorithms and stay far ahead of would-be competitors. The rich get richer.

Google is not doing this as a public service. It uses this personalized information to place targeted ads and match companies with the people they are trying to reach and influence. As they say, when you use a search engine, you are not the customer, you are the product.

In 2018, Google had $136 billion in revenue, of which $116 billion came from advertising. Google's ad revenue has generated an enormous cash flow, which it invests in other projects, including Gmail; Google Chrome (which has displaced Microsoft's Internet Explorer as the most popular browser); Google Docs, Google Sheets, and Google Slides (which threaten Microsoft Word, Excel, and PowerPoint); Google Maps; and of course, Google cars.

There are two components to Google's success. The first is its ability to figure out what people are looking for when they type in cryptic phrases with jumbled, misspelled words. The second is its ability to determine which web pages are the most relevant for the intended search.

Part of Google's advantage is that it can provide custom, personalized searches, using data from a user's previous searches, as well as information from Gmail messages and Google+ profiles to select the most relevant sites. However, it has another not-so-secret weapon: its extensive use of A/B tests.

Understandably, Google does not publicize the details of its algorithms, but we do know that it does lots of A/B testing. It even offers customers a free A/B testing tool, called Google Optimize.

An A/B test is like a laboratory experiment. Two versions of a page are created—typically the current page and a proposed modification, perhaps with a different banner, headline, product description, testimonial, or button label. A random number generator is used to determine which page a user sees. In the language of a medical test, those sent to the current page are the *control group* and those sent to the modified page are the *treatment group*. A pre-specified metric, like purchases or mailing-list signups—is used to determine the winning page. There can be more than two alternatives (an A/B/C/D, etc. test) and there can be multiple differences between pages.

Google's first A/B test, conducted in 2000, was used to determine the optimal number of search results to display on a page. They decided that ten was best, and that is still the number used today.

Dan Siroker left Google in 2008 to become the Director of Analytics for Barack Obama's 2008 presidential campaign. One of his projects involved the design of the splash page on the campaign's main website. Figure 9.2 shows the original design before Siroker ran A/B tests.

The metric of interest was the frequency with which people hit the "SIGN UP" button that gave the campaign an e-mail address that could be used for future communications and solicitations. Siroker ran A/B tests on four different buttons and six different media. They found that replacing the original "SIGN UP" button with a "LEARN MORE" button increased the sign-up rate by 18.6 percent, and that replacing the original color photo of Obama with a black-and-white family photo increased the sign-up rate by 13.1 percent. The two changes together, shown in Figure 9.3, increased the signup rate by 40.6 percent, from 8.26 percent to 11.60 percent, which represented three million additional e-mail addresses.

Google uses A/B tests to help them turn search queries into meaningful words and phrases, decide what users are searching for, identify the web pages that searchers will find most useful, and determine how to display results.

Google runs tens of thousands, perhaps hundreds of thousands, of A/B tests every year in a virtuous cycle of searchers being used as free labor to improve the search results and lock them in as Google users.

The next time you do a Google search, you might well be participating in an A/B test.

Figure 9.2 The original Obama 2008 splash page.
Dan Siroker

Figure 9.3 The new-and-improved Obama 2008 splash page.
Optimizely

Controlled Experiments

Franklin D. Roosevelt was President of the United States for twelve years, from March 1933 to April 1945—from the depths of the Great Depression through the defeat of Nazi Germany in the Second World War. His inspirational speeches, innovative social programs, and reassuring radio "fireside chats" helped create the idea that the federal government, in general, and the U.S. President, in particular, were responsible for the country's well-being.

Though he projected an image of strength and optimism, Roosevelt had been stricken with an illness in 1921, when he was thirty-nine years old, that left him permanently paralyzed from the waist down. His public appearances were carefully managed to prevent the public from seeing his wheelchair. In public speeches, he stood upright, supported by aides or by a tight grip on a strong lectern.

Roosevelt's disease was diagnosed as poliomyelitis (polio), which was a recurring epidemic at the time, though it is now believed that it is more likely that he had Guillain–Barré syndrome. In 1938 Roosevelt founded the National Foundation for Infantile Paralysis, now known as the March of Dimes, to support polio research and education. Fittingly, after Roosevelt's death, the federal government replaced the Winged Liberty Head dime with the Roosevelt dime on January 30, 1946, which would have been Roosevelt's sixty-fourth birthday.

Polio is an acute viral disease that causes paralysis, muscular atrophy, permanent deformities, and even death. It is an especially cruel disease in that most of its victims are children. The U.S. had its first polio epidemic in 1916 and, over the next forty years, hundreds of thousands of Americans were afflicted. In the early 1950s, more than 30,000 cases of acute poliomyelitis were reported each year.

Researchers, many supported by the March of Dimes, found that most adults had experienced a mild polio infection during their lives, with their bodies producing antibodies that not only warded off the infection but made their bodies immune to another attack. Similarly, they found that polio was rarest in societies with the poorest hygiene. The explanation is that almost all the children in these societies were exposed to the contagious virus while still young enough to be protected by their mother's antibodies and, so, they developed their own antibodies without ever suffering from the disease.

Scientists consequently worked to develop a safe vaccine that would provoke the body to develop antibodies, without causing paralysis or worse. A few vaccines had been tried in the 1930s and then abandoned because they sometimes caused the disease that they had been designed to prevent. By the 1950s, extensive laboratory work had turned up several promising vaccines that seemed to produce safe antibodies against polio.

In 1954, the Public Health Service organized a nationwide test of Jonas Salk's polio vaccine, involving two million schoolchildren. Because polio epidemics varied greatly from place to place and from year to year throughout the 1940s and early 1950s, the Public Health Service decided not to offer the vaccine to all children, either nationwide or in a particular city. Otherwise, there would have been no control group and the Health Service would not have been able to tell if variations in polio incidence were due to the vaccine or to the vagaries of epidemics. Instead, it was proposed that the vaccine be offered to all second graders at selected schools. The experience of these children (the treatment group) could then be compared to the school's first and third graders (the control group) who were not offered the vaccine. Ideally, the treatment group and the control group would be alike in all respects, but for the fact that the treatment group received the vaccine and the control group did not.

There were two problems with this proposed experiment. First, participation was voluntary and it was feared that those who agreed to be vaccinated would tend to have higher incomes and better hygiene and, as explained earlier, be more susceptible to polio. Second, school doctors were instructed to look for both acute and mild cases of polio, and the mild cases were not easily diagnosed. If doctors knew that many second graders were vaccinated, while first and third graders were not, this knowledge might influence the diagnosis. A doctor who hoped that the vaccine would be successful might be more apt to see polio symptoms in the unvaccinated than in the vaccinated.

A second proposal was to run a double-blind test, in which only half of the volunteer children would be given the Salk vaccine, and neither the children nor the doctors would know whether the child received the vaccine or a placebo solution of salt and water. Each child's injection fluid would be chosen randomly from a box and the serial number recorded. Only after the incidence of polio had been diagnosed, would it be revealed whether the child received vaccine or placebo. The primary objection to this proposal was the awkwardness of asking parents to support a program in which there was only a fifty percent chance that their children would be vaccinated.

As it turned out, about half of the schools decided to inoculate all second-grade volunteers and use the first and third graders as a control group. The remaining half agreed to a double-blind test using the placebo children as a control group. The results are in Table 9.1. The Salk vaccine reduced the incidence of polio in both cases. The number of diagnosed polio cases fell by about fifty-four percent in the first approach and by sixty-one percent with the placebo control group. If the double-blind experiment had not been conducted, the case for the Salk vaccine would have been less convincing because the decline was smaller and because skeptics might have attributed this decline to a subconscious desire by doctors to have the vaccine work.

Table 9.1 *The results of the 1954 nationwide test of a polio vaccine.*

	First- and Third-Grade Control Group		Double-Blind with Placebo	
	Children	Polio per 100,000	Children	Polio per 100,000
Treatment group	221,998	25	200,745	28
Control group	725,173	54	201,229	71
No consent	123,605	44	338,778	46

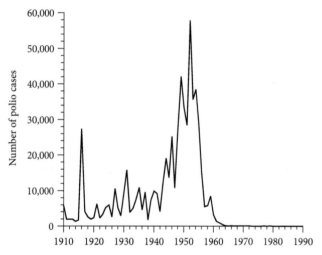

Figure 9.4 The disappearance of Polio.

The 1954 tests were a landmark national public health experiment that provided convincing evidence of the value of a polio vaccine. The Salk vaccine was eventually replaced by the even safer and more effective Sabin preparation. Figure 9.4 shows the dramatic decline in polio cases after national immunization programs began in the 1950s. Today, about ninety-two percent of all children in the U.S. receive polio vaccinations between the ages of nineteen and thirty-six months and no polio cases have originated in the U.S. since 1979.

Natural Experiments

It can be difficult to draw conclusions from observational data (things we observe as opposed to controlled experiments). For example, people who serve in the U.S. military tend to have lower incomes after they leave the military, compared to their peers who were not in the military. This might be because the training that people receive in the military is less useful for civilian jobs than the education and work experience they might have received if they had not enlisted. Or, it might be due to a self-selection bias that is endemic in observational data. The different outcomes for people who choose an activity and those who don't may be due to differences among the people who make such choices. People with college degrees may differ from people without college degrees, not because of what they learn in college, but because people who choose to go to college and complete their degree are different from those who don't. Married people may differ from the unmarried not because of what marriage does to them, but because of differences between those who choose to marry and those who choose not to. People who choose to enlist in the military may have relatively low pay afterward because they had limited job prospects and no interest in going to college when they chose to enlist.

The way to get around the problems with observational data is to run a randomized controlled trial, but we can't very well force randomly selected people to marry or not marry, or go to college or not go. However, we can force people to enlist in the military. A natural experiment occurred during the Vietnam War when the U.S. Selective Service system used a lottery to determine randomly selected people who were drafted or not drafted.

The first lottery was held on December 1, 1969, and determined the draft order of 850,000 men born between January 1, 1944, and December 31, 1950. The 366 days of the year (including February 29) were written on slips of paper and inserted into blue plastic capsules that were poured into a large plastic container that looked like the bottom half of a water-cooler bottle.

Anxious men (and their families) were told that the first 125 birthdates selected were very likely be drafted; numbers 126–250 might not be drafted, and numbers higher than 250 were safe.

In a nationally televised ceremony, Alexander Pirnie, the top Republican on the House Armed Services Committee, pulled out the capsule containing the date September 14, and this date was placed on a board next to the number 001, indicating that men born on September 14 would be the first to be called to duty.

The remaining picks were made by a group of young people in order to demonstrate that young people were part of the process. Paul Murray, a student from Rhode Island, made the second pick (April 24), and so it went until the last day selected (June 8) was given the number 366. As it turned out, all dates with numbers 195 or lower were required to report for service.

It was dramatic and anxiety producing, but this draft lottery did create a natural experiment that avoided self-selection bias. Not everyone with a low number was inducted (some had health problems such as bone spurs), but the biggest difference between those with the first 195 birth dates and those with later birth dates is that the first group was eligible for induction.

A subsequent study found that there was no difference in the average earnings of draft-eligible and draft-ineligible white males before the lottery, but a substantial difference afterward. Being drafted mattered, and not for the better. Those who got low draft numbers had lower incomes, on average, than those who avoided being selected. The income differences were naturally largest during the time that the draft-eligible males were in the military, but continued after they fulfilled their military obligations. Ten years after their service, veterans earned, on average, fifteen percent less than non-veterans. Military experience was a poor substitute for civilian education and work experience. The differences for non-white males were less pronounced, but all veterans generally sacrificed some of their future financial well-being as well as their time.

Dr. Spock's Overlooked Women

In 1946 a pediatrician named Benjamin Spock published *The Common Sense Book of Baby and Child Care*, which sold more than fifty million copies worldwide. When he died in 1998 at age ninety-six, a *Time* magazine obituary said that Dr. Spock "singlehandedly changed the way parents raise their children."

Before Spock, the conventional wisdom was that children need strict discipline—fixed timetables for eating and for sleeping, and spankings for misbehavior. Spock advised parents to "trust your own common sense." If a baby is not hungry, don't make him eat. If a baby is not tired, don't make her sleep. Above all, babies need love, not rules and punishment. This flexibility was mocked by some, but embraced by many. Conversations among parents often included the phrase, "According to Dr. Spock,...". Even, in 2019, seventy-three years after the publication of *Baby and Child Care*, an article in *The Washington Post* on Donald Trump's presidency said, "And, according to Dr. Spock,...".

Spock was a vocal opponent of the Vietnam War and some blamed his child-rearing advice for the rebellious youth with shabby clothes and shaggy hair who protested the war. In 1968 Spock and four other defendants were tried in a Massachusetts federal court for conspiracy to violate the Military Service Act of 1967 by counseling, aiding, and abetting resistance to the draft. Spock and three other defendants were convicted, but appealed.

The court clerk testified that he selected potential jurors by putting his finger randomly on a list of adult residents of the district. The people he selected were sent questionnaires and, after those disqualified by statute were eliminated, batches of 300 were called to a central jury box, from which a panel of 100 (called a "venire") was selected. As it turned out, even though more than half of the residents in the district were female, only nine of the 100 people on Dr. Spock's venire were women.

The jury selection process may have been biased. A finger placed blindly on a list will almost certainly be close to several names, and the court clerk may have favored males when deciding which names to use. Under cross examination, the clerk admitted that this may have been the case:

Answer: ...I put my finger on the place and on a name on the page and then I make a mark next to it with a pen.

Question: Do you do that by not looking at the page?

Answer: I have to look at it enough to know where it is in relation to my finger.

Question: Yes.

Answer: I do not intend to look carefully at the name....

Question: I assume that at some point you have to look at the name in order to send out the questionnaire?

Answer: Correct.

Question: Do you have any explanation for that [the disparity between the number of questionnaires sent to the men and women] except the possibility that you might have seen the name and recognized it as a woman's name and figured it is a little more efficient not to send out too many questionnaires to women?

Answer: That is the only possible explanation other than pure chance...

The court records do not reveal how the panel of 100 was selected from the central jury box, but this, too, may have been biased by subjective factors.

These were observational data, but the analysis was motivated by the plausible theory that women may have been discriminated against—as opposed to looking at the data in 100 different ways and finding that there were an unusual number of left-handed people whose last names began with the letter S. A randomized controlled trial was not possible, but there was a natural control group.

Several months after the end of the trial, the defense obtained data showing that out of 598 jurors recently used by Dr. Spock's trial judge, only eighty-seven (14.6 percent) were female, while twenty-nine percent of the 2,378 jurors used by the other six judges in the district were female. To suggest how this bias may have prejudiced Spock's chances of acquittal, the appeal cited a 1968 Gallup poll in which fifty percent of the males labeled themselves as hawks and thirty-three percent doves on Vietnam, as compared to thirty-two percent hawks and forty-nine percent doves among females.

It is also conceivable that women who raised their children "according to Dr. Spock" might have been sympathetic to his anti-war activities. The appeal argued that the probability that randomly selected juries would be so disproportionately male was small and that, "The conclusion, therefore, is virtually inescapable that the clerk must have drawn the venires for the trial judge from the central jury box in a fashion that somehow systematically reduced the proportion of women jurors." (The probability of such a large gender disparity between the jurors used by this judge and the jurors used by the other six judges is about one in twenty-eight trillion).

Spock's conviction was eventually overturned, though on First Amendment grounds rather than because of a flawed jury selection. However, a law passed in 1969 mandated the random selection of juries in federal courts using statistically accepted techniques, such as random-number generators. Subsequently, males and females have been equally represented on venires in federal courts.

Theory Before Data

Many frivolous, even ludicrous, strategies for beating the stock market have been uncovered by ransacking historical data, looking for patterns. Stock market patterns are particularly seductive because they dangle the lure of easy money—indeed, money for nothing. The silly systems mentioned at the start of this chapter are just a small sample. Financial astrologists study the positions of planets and stars. The Sports Illustrated Swimsuit indicator is based on whether the cover model is from the U.S. The headache systems is based on aspirin sales. The BB system is based on butter production in Bangladesh.

Viable strategies begin with a plausible theory—a logical basis—instead of unconstrained data exploration. Daniel Kahneman and Amos Tversky documented a type of fallacious reasoning they called "the law of small numbers." When something unusual happens, we tend to leap to the easy conclusion that it is likely to happen again—that the event is typical, not unusual. This can be the basis for an over-reaction to trivial events. We might think that a basketball player who makes a difficult shot is a good shooter. We might think that a commentator who makes a correct political prediction is astute. We might think that a person who tells a funny joke is a natural comedian. Kahneman and Tversky collected a variety of formal experimental evidence that confirmed their hypothesis that people tend to overweight new information.

In the stock market, Keynes observed that "day-to-day fluctuations in the profits of existing investments, which are obviously of an ephemeral and nonsignificant character, tend to have an altogether excessive, and even absurd, influence on the [stock] market." If true, such over-reaction might be the basis for Warren Buffett's memorable advice, "Be fearful when others are greedy, and be greedy when others are fearful." If investors often over react, causing excessive fluctuations in stock prices, it may be profitable to bet that large price movements will be followed by price reversals.

One example (from an essentially endless list) involved Oracle, a software powerhouse, on December 9, 1997. Analysts had been expecting Oracle's second-quarter sales to be thirty-five percent higher than a year earlier and its profits to be twenty-five percent higher. After the market closed on December 8, Oracle reported that its second-quarter sales were only twenty-three percent higher than a year earlier and its profits were only four percent higher. The next day, 171.8 million Oracle shares were traded, more than one-sixth of all Oracle shares outstanding, and the stock's price fell twenty-nine percent, reducing Oracle's total market value by more than $9 billion.

As is so often the case, the market over-reacted. The annual return on Oracle stock over the next twenty-one years, through December 31, 2018, was 15.8 percent, compared to 4.1 percent for the S&P 500. A $10,000 investment in Oracle the day after its 1997 crash would have grown to $133,000, compared to $39,000 for the S&P 500.

More recently, on January 24, 2013, Apple reported a record quarterly profit of $13.1 billion, selling twenty-eight percent more iPhones and forty-eight percent more iPads than a year earlier, but the stock dropped more than twelve percent, reducing its market value by $50 billion. Apple had sold a record 47.8 million iPhones, but this was less than the consensus forecast of fifty million. Earnings per share were higher than predicted ($13.81 versus $13.44), and so was revenue ($54.7 billion versus $54.5 billion), but investors were used to Apple clobbering forecasts. A bit of a paradox here. If analysts expected Apple to beat their forecasts, why didn't they raise their forecasts? In any case, investors were scared and the market over-reacted. Figure 9.5 shows that, from January 24, 2013 through December 31, 2019, the S&P 500 is up about 100 percent, while Apple is up 400 percent.

Are these examples of selective recall or evidence of a commonplace over-reaction in the stock market? To find out, Gary analyzed daily returns for the stocks in the Dow Jones Industrial Average from October 1, 1928, when the Dow was expanded from twenty to thirty stocks, through December 31, 2015, a total of 22,965 trading days.

Every day, each stock's *adjusted* daily return was calculated relative to the average return on the other twenty-nine Dow stocks that day. The use

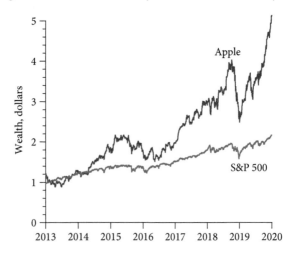

Figure 9.5 Go Apple.

of adjusted returns is intended to focus attention on stocks that are buoyed or depressed by idiosyncratic news or opinions specific to an individual company, as opposed to general market surges or crashes caused by macro-economic news or emotions.

A big day for a Dow stock was defined as a day when its adjusted return was above five percent or below negative five percent. For a robustness check, Gary also redid did the calculations with big-day cutoffs of six, seven, eight, nine, or ten percent. These alternative cutoffs did not affect the conclusions.

The performance of each stock that experienced a big day was tracked for ten days after the big day. Table 9.2 shows that most stocks went down after they had a big up day and went up after they had a big down day. This difference was highly statistically significant.

Figure 9.6 and Figure 9.7 show the average adjusted daily returns and the average adjusted cumulative returns for the ten days following positive and negative big days. The average returns were negative for the first nine days following a positive big day and positive for nine of the ten days following a negative big day. The differences between the cumulative returns after positive and negative big days were substantial and highly statistically significant. On day ten, there was an average 0.59 percent cumulative loss following a positive big day and an average 0.87 percent cumulative gain following a negative big day (representing very large annualized returns).

The over-reaction seems more excessive for negative events in that the return reversals are larger following big down days than big up days. Perhaps this has something to do with the loss aversion documented by Kahneman and Tversky. Over-reaction is more pronounced for bad

Table 9.2 *Percentage of stocks with positive adjusted returns the day after a big day.*

Cutoff, Percent	After Big Up Day	After Big Down Day
5	44.34	53.60
6	44.34	52.45
7	43.95	50.90
8	42.87	53.83
9	45.35	53.22
10	46.10	53.12

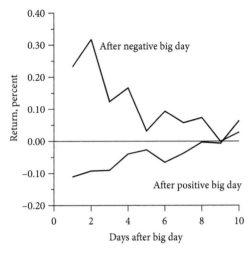

Figure 9.6 Average daily adjusted return after a five percent big day.

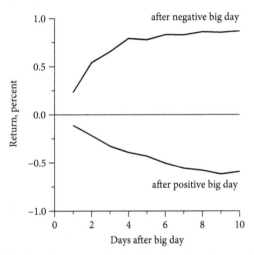

Figure 9.7 Cumulative average daily adjusted return after a five percent big day.

news than for good news, because investors are more susceptible to panic than greed.

This study obtained interesting, plausible, and potentially useful results because it was motivated by a plausible theory—that investors over-react to news events—instead of using a data mining algorithm to find a puzzling and fleeting pattern, like stock prices tending to rise fourteen days after they had fallen two days in a row and then risen two days in a row.

Exuberant Expectations

Bonds pay interest and stocks pay dividends, which is why Warren Buffett calls stocks "disguised bonds." Up until the late 1950s, many investors gauged stocks and bonds by comparing the interest rates on bonds with the dividend yields on stocks. A $100 bond that pays $3 in annual interest has a three percent annual return. A $100 stock that pays $3 in annual dividends has a three percent dividend yield. Since stocks are riskier than bonds, investors used to believe that dividend yields should be higher than interest rates.

Figure 9.8 shows that, up until 1958, the average dividend yield for the S&P 500 index of stock prices almost always exceeded the interest rate on

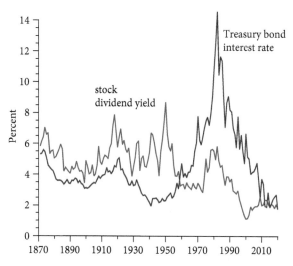

Figure 9.8 Dividend yields and interest rates.

Treasury bonds. In 1950, the average dividend yield was nearly nine percent, while the interest rate on Treasury bonds was only two percent.

This comparison was misguided. Indeed, Buffett has argued that there is an important difference between interest and dividends—which is why he seldom invests in bonds. Bond interest payments are fixed, but stock dividends can (and usually do) increase over time, along with corporate earnings. If a stock has a nine percent dividend yield and the dividend grows by five percent a year, the annual rate of return is effectively fourteen percent, not nine percent.

As the value of growth was increasingly recognized in the 1950s and 1960s, rising stock prices pushed dividend yields below bond yields. By the early 1970s, investors seemed to be interested *only* in growth, especially the premier growth stocks labeled the "Nifty 50." Among these select few were IBM, Xerox, Disney, McDonald's, Avon, Polaroid, and Schlumberger.

Investors swung from one extreme to another, as they often do, paying what, in retrospect, were ridiculous prices. They went from ignoring growth to fixating on growth. The average price-earnings (P/E) ratio for the S&P 500 since 1871 has been about 15. In 1972, sixteen of the Nifty 50 stocks had P/Es above 50; some even topped 100.

David Dreman recounted the dreams of missed opportunities that danced in investors' heads:

Had someone put $10,000 in Haloid-Xerox in 1960, the year the first plain copier, the 914, was introduced, the investment would have been worth $16.5 million a decade later. McDonald's earnings increased several thousand times in the 1961–66 period, and then, more demurely, quadrupled again by 1971, the year of its eight billionth hamburger. Anyone astute enough to buy McDonald's stock in 1965, when it went public, would have made fortyfold his money in the next seven years. An investor who plunked $2,750 into Thomas J. Watson's Computing and Tabulating Company in 1914 would have had over $20 million in IBM stock by the beginning of the 1970s.

The unfortunate consequence of this fixation on the Nifty 50 was the belief that there is never a bad time to buy a growth stock, nor is there too high a price to pay. A money manager infatuated with growth stocks wrote that, "The time to buy a growth stock is now. The whole purpose in such an investment is to participate in future larger earnings, so *ipso facto* any delay in making the commitment is defeating." He didn't mention price, an omission that is invariably an investing sin.

The subsequent performance of many of the Nifty 50 was disappointing. In 1973 Avon sold for $140 a share, sixty times earnings. In 1974, the price collapsed to $19; twelve years later, in 1986, it was selling for $25 a share. Polaroid sold for $150 in 1972 (115 times earnings!), $14 in 1974, and $42 in 1986, before going bankrupt in 2001.

These were not isolated cases. Many glamour stocks that were pushed to extraordinary P/E levels in the 1970s did substantially worse than the market over the next several decades. An investor with exquisitely bad timing who bought the Nifty 50 stock with P/Es above 50 at the end of 1972 would have had an average annual return of negative fourteen percent over the next forty-five years (through December 2018), compared to an average return of more than ten percent for the S&P 500.

How to Avoid Being Misled by Phantom Patterns

Patterns need not be combinations of numbers. For example, employees—ranging from clerks to CEOs—who do their jobs extremely well are often less successful when they are promoted to new positions—a disappointment immortalized by the Peter Principle: "managers rise to the level of their incompetence."

Patterns in observational data can be misleading because of self-selection bias, in that observed differences among people making different choices may be due to the type of people making such choices. Randomized controlled trials and A/B tests, when practical, are effective ways of determining cause and effect. When forced to use observational data, it is important that the theories that are tested are specified before looking at the data. Otherwise, we are likely to be fooled by phantom patterns.

All About That Bayes

I n the eighteenth century, a Presbyterian minister named Thomas Bayes wrestled with a daunting question—the probability that God exists. Reverend Bayes was not only a minister, he had studied logic and, most likely, mathematics at the University of Edinburgh (Presbyterians were not allowed to attend English universities).

Many intellectuals at the time debated what is still a troubling logical conundrum. If a benevolent god created the universe and keeps a watchful eye on us, why is there so much misery and evil in the world? Would a caring god allow good people to starve, the faithful to die of horrible diseases, and innocent children to be slaughtered by soldiers?

Bayes's initial foray into this debate was a high-minded, but not very convincing, pamphlet published under the pseudonym James Noon, with a seemingly endless title: *Divine benevolence: or, An attempt to prove that the principal end of the divine providence and government is the happiness of his creatures. Being an answer to a pamphlet, entitled, Divine rectitude [by John Balguy] . . . With a refutation of the notions therein advanced concerning beauty and order, the reason of punishment, and the necessity of a state of trial antecedent to perfect happiness.*

Justifiably unsatisfied with back-and-forth arguments, Bayes set out to provide a mathematical argument—a calculation of the probability of god's existence!

At the time, mathematicians were rapidly developing theorems and proofs for calculating the probability of various games of chance. For example, if a fair coin is tossed ten times, what is the probability of five heads and five tails? (The answer is 0.246.)

Bayes posed the reverse question, the *inverse* probability: if a coin is flipped ten times and lands heads five times and tails five times, what is the probability that it is a fair coin? If he could answer such a question, then perhaps he could calculate the probability of god's existence.

Instead of trying to determine this probability:

If there is a god, what if the probability the world would be the way it is?

Bayes wanted to calculate the inverse probability:

If the world is the way it is, what is the probability that there is a god?

Bayes was not able to calculate the probability of God's existence, but his work on how to go from one probability to its inverse has turned out to be incredibly valuable and is now the foundation for the Bayesian approach to probability and statistics.

Bayes' Theorem

In calculating inverse probabilities for various games of chance, Bayes discovered that the calculation requires a *prior probability* (before the outcome is observed), which is then revised to a *posterior probability* (after the outcome is observed).

Let's work through a simple example to see how prior probabilities are revised into posterior probabilities. Suppose that twenty coins are put in a black bag; nineteen are regular coins and one coin has heads on both sides. A coin is randomly picked from the bag and one side is shown to be heads. What is the probability that it is the two-headed coin? What's your guess? A ten percent chance that it is the two-headed coin? Thirty percent? Fifty percent? Seventy percent?

Table E.1 shows a straightforward way of organizing our thinking. Suppose the experiment is done forty times. Since there is a one in twenty chance of selecting the two-headed coin, we can expect to pick the two-headed coin

Table E.1 *An inverse probability.*

	Heads is Shown	Tails is Shown	Total
Two-headed coin is picked	2	0	2
Normal coin is picked	19	19	38
Total	21	19	40

twice and a normal coin thirty-eight times. On those two occasions when the two-headed coin is picked, the side that is revealed will always be heads. For the thirty-eight occasions when a normal coin is selected, we can expect to see the heads side nineteen times and the tails side nineteen times.

So, out of the twenty-one times that the revealed side is heads, there is a 9.5 percent chance that it is the two-headed coin, $2/21 = 0.095$.

Before the coin is chosen, the prior probability that the two-headed coin will be selected is $1/20$. After the coin is chosen and one side is shown to be a head, the posterior probability that it is the two-headed coin nearly doubles, from $1/20$ to $2/21$.

This insight is critical for the modern usage of Bayes' rule. The posterior probability revises the prior probability based on data that have become available. This is the essence of learning from data—using new information to revise our opinion.

John Maynard Keynes is credited with a wonderful rejoinder to the criticism that he had changed his policy recommendations: "When my information changes, I alter my conclusions. What do you do, sir?" It would be intellectually dishonest not to change our opinion in light of new information. Bayes' rule guides us in such revisions.

Let's apply this reasoning to a medical diagnosis.

Medical Diagnoses

Some medical diagnoses are straightforward. Yes, you twisted your ankle. Yes, you hit your thumb with a hammer. Other diagnoses are less certain. Suppose that during a routine medical examination, a doctor finds a suspicious lump in a female patient's breast. There is no way of knowing for certain if the lump is malignant or benign. Perhaps the doctor knows that, among comparable women with similar medical records, the lump turns out to be malignant in one out of 100 cases. To gather more information, the doctor orders a mammogram X-ray. It is known that in those cases where the lump is malignant, there is a 0.80 probability that the X-ray will give a positive reading and, when the lump is benign, there is a 0.90 probability that the X-ray will give a negative reading.

The prior probability that the lump is malignant is 0.01; the posterior probability is the revised probability based on the X-ray result. Suppose the X-ray comes back positive. What is your guesstimate of the posterior probability that the lump is malignant? Eighty percent? Ninety percent?

Using Bayes' rule, it turns out that the posterior probability is 7.5 percent. This is a particularly nice example because it is a very striking illustration of how probabilities and inverse probabilities can be quite different. Even though a malignant tumor will be correctly identified as such eighty percent of the time, there is only a 7.5 percent chance that a positive X-ray reflects a malignant tumor.

It also shows how Bayes' rule modifies probabilities in the light of new data—here, from the 0.01 prior probability (before the X-ray) to a 0.075 posterior probability (after the X-ray). If the X-ray had come back negative, the posterior probability would have been revised downward to 0.002.

Bayes' rule can be used over and over to continue revising our probability based on the results of additional X-rays. Suppose that the first X-ray comes back positive and the doctor orders a second X-ray. Now Bayes' rule tells us that the 0.075 probability after the first test is either increased to 0.393 or reduced to 0.018, depending on the second X-ray reading.

It should be no surprise that computers rely heavily on Bayes' rule when they use data to revise the probability that you will like a certain movie, gadget, or political position. For example, chances are that if you loved the television show *Breaking Bad*, you will like the spinoffs *Better Call Saul* and *El Camino*. Bayes' rule can be used to quantify those probabilities. Not only that, Bayes' rule can take into account other relevant information that might affect your preferences, such as your age, gender, occupation, and income, and it can update its probability estimate as it learns more about you—the same way that Bayes' rule took into account the results of a second X-ray in estimating a malignancy probability.

In the use of Bayes' rule, computers have a clear advantage over humans because we tend to make two systematic mistakes. First, we don't appreciate fully the difference between probabilities and inverse probabilities and, second, we don't revise probabilities the way Bayes' rule says we should.

In the mammogram example, what was your guesstimate of the probability that the lump was malignant after the first X-ray came back positive? Was it closer to 7.5 percent or eighty percent? One hundred doctors were asked the same question and ninety-five of the doctors gave probabilities of around seventy-five percent, even though the correct probability is one-tenth that! According to the researcher who conducted this survey, "The erring physicians usually report that they assumed that the probability of cancer given that the patient has a positive X-ray … was approximately equal to the probability of a positive X-ray in a patient with cancer." Medical misinterpretations of inverse probabilities can have tragic consequences.

In addition to confusing probabilities and inverse probabilities, humans are not very good at revising probabilities based on new data. Remember the example of twenty coins, one of which is two-headed? What was your guess of the probability that the selected coin was two-headed? Was it ten percent (the correct answer), or closer to thirty, fifty, or even seventy percent? Don't be embarrassed. Human intuition is notoriously imperfect when it comes to revising probabilities in the light of new information.

This flaw is so common, it even has a name: the *base rate fallacy*. When people have general information and specific data, they tend to emphasize the specific data and neglect the general (or base rate) information. Suppose the base rate is that less than one-quarter of one percent of a nation's women are lawyers, and the specific data are that Jill is wearing a business suit and carrying a briefcase. What is the probability that Jill is a lawyer? Most people fixate on the pattern that lawyers often wear business suits and carry briefcases and pay insufficient attention to the 0.25 percent base rate.

This is yet another way in which we are susceptible to being fooled by phantom patterns. We are tempted to attach a great deal of importance to a small bit of data—a memorable or unusual pattern—when we shouldn't.

Bayes' Rule in the Courtroom

Jury trials also give convincing evidence of human confusion about inverse probabilities and human errors in revising probabilities. For example, a law review article gave this example of how jurors often err in assessing the value of associative evidence, such as fingerprints or blood samples that match those of the accused.

Imagine that you are a juror at a murder trial and, based on the evidence that you have heard so far, you believe that it is fifty-fifty whether or not the defendant is guilty. The prosecution now presents evidence showing conclusively that, based on the angle of the murder blow, the assailant is right-handed. You look at the defense table and see the defendant taking notes with his right hand. What is your revised probability that the defendant is guilty?

According to Bayes' rule, the posterior probability of guilt is 0.53, which is only a smidgen above the 0.50 prior probability. The right-handed testimony has little probative value because right-handedness is so common in the population. It is worth little more than evidence that the assailant has a nose. Yet, the authors of the law review article argued that, in their experience, jurors often attach great significance to evidence that has little or no value.

The situation would be different if the defendant happened to be left-handed and conclusive testimony were presented that the assailant is left-handed, since left-handedness is uncommon. In that situation, the posterior probability jumps to 0.91.

Oddly enough, while jurors often overweight trivial evidence, they typically underweight substantial evidence. Researchers have conducted mock trials where volunteer jurors hear evidence and are asked to record their personal probabilities of the accused person's guilt. In one experiment, 144 volunteer jurors were told that a liquor store had been robbed by a man wearing a ski mask. The police arrested a suspect near the store whose height, weight, and clothing matched the clerk's description. The ski mask and money were found in a nearby trash can.

After hearing this evidence, the jurors were asked to write down their estimate of the probability that the arrested man "really did it." The average probability was 0.25.

Then a forensic expert testified that samples of the suspect's hair matched a hair found inside the ski mask and that only two percent of the population has hair matching that in the ski mask. Half of the jurors were also told that in a city of one million people, this two percent probability meant that approximately 20,000 people would have a hair match.

Nineteen of the jurors gave a posterior probability of 0.98, evidently believing that if only two percent of the innocent people have this hair type, there is only a two percent probability that the accused person is innocent. This is called the Prosecutor's Fallacy because it is an argument used by prosecutors to secure a conviction. It is a fallacy because the jurors are confusing probabilities and inverse probabilities—specifically, they do not distinguish between the probability that an innocent percent has this hair type with the inverse probability that a person with this hair type is innocent.

Six of the seventy-two jurors who were given only the two percent probability and twelve of the seventy-two jurors who were given both the two percent probability and the 20,000 number did not revise their probabilities at all, evidently believing that the hair match was useless information because so many people have this hair type. This is called the

Defense Attorney's Fallacy because it is an argument used by defense attorneys. It is a fallacy because the defendant was not randomly selected from 20,000 people with this hair type. The other evidence regarding the suspect's description and the discarded ski mask and money had already persuaded the jurors that there was a significant probably that the accused person is guilty.

For a prior probability of 0.25, Bayes' rule implies that the posterior probability, after taking the hair match into account, is 0.94. Three-fourths of the jurors did not fall for either the Prosecutor or Defense Attorney fallacy; however, their revised probabilities were consistently more conservative than the Bayesian posterior probabilities implied by their priors. The average revised probability was only 0.63.

This conservatism has shown up again and again in mock jury experiments. In ten such experiments, with average Bayesian posteriors ranging from 0.80 to 0.997, the average jurors' revised probabilities were consistently lower, ranging from 0.28 to 0.75.

Jurors, who are human after all, tend to believe that trivial data are important and important data are not persuasive. They think a flukey pattern is important, while a boatload of data is not.

One extreme example was O. J. Simpson's "trial of the century," in which the jurors seemingly concluded that the DNA evidence pointing to guilt was less important than the fact that blood-soaked leather gloves found at the crime scene did not fit comfortably on Simpson's hands: "If it doesn't fit, you must acquit." There were many possible explanations for the poor fit: Simpson was wearing latex gloves when he tried on the leather gloves; the gloves may have become stiff from having been soaked in blood and frozen and unfrozen several times; Simpson's hands were swollen because he had stopped taking his arthritis medicine two weeks earlier; and Simpson pretended that the gloves were difficult to put on.

The jurors valued the trivial over the serious, perhaps because they had trouble assessing the importance of each, or perhaps they saw what they wanted to see.

It is ironic that Bayesian analyses, which are designed to allow humans to express and revise their personal probabilities, are done better by computers than by humans. On the other hand, computers still have an insurmountable problem: assessing whether specific data *should* be used to revise probabilities—distinguishing good data from bad data. Computers are terrible at that.

Humans have the clear advantage when it comes to assessing the quality of the data. Think again about the mock trial involving a robber wearing a ski mask, and the expert testimony that a hair found in the ski mask is found in only two percent of the population. Perhaps the people who did not change their guilty probabilities or did not change them as much as implied by Bayes' rule didn't trust the expert testimony. They may well have discounted the two percent figure because they assumed that the prosecution would use the paid expert who gave the most favorable testimony. Or they may have considered the possibility that the police planted the hair in the ski mask. Their revised probabilities may have been the right subjective probabilities for them.

Jay was recently summoned for jury duty and sat in the gallery for four full days while prospective jurors were called up in groups of twelve and interrogated by the prosecution and defense attorneys in the hopes of finding twelve suitable jurors. There were eighty potential jurors at the beginning and the attorneys went through seventy-five of them before they settled on twelve jurors. Remarkably, Jay was among the five who were not interrogated.

He was a bit relieved, because there was bound to be a lively discussion between him and the defendant's attorney, who had repeatedly instructed the jury on how they should handle "propensity evidence." In some situations, judges may allow the prosecution to present evidence of similar crimes in the defendant's past. The jury was told by the attorney that, "If the evidence *in this case* is not proven beyond a reasonable doubt, you cannot then say 'well, he's done something similar in the past. Where there's smoke, there's fire', and then decide he's guilty. The People must still prove each charge *in this case* beyond a reasonable doubt."

Evidence of past crimes is not sufficient, by itself, to prove that the defendant is guilty of the current crime, but the attorney seemed to be arguing that evidence of past crimes should never persuade jurors to change their conclusion from "not guilty" to "guilty." If that were true, then propensity evidence would be given no weight whatsoever, and judges should never allow the prosecution to introduce evidence of similar crimes by the defendant.

Suppose that a juror considers a ninety-five percent probability of guilt as "beyond a reasonable doubt" and all of the evidence related to the current crime puts the probability of guilt at ninety-four percent. Evidence of similar crimes committed in the past would surely push the probability above the ninety-five percent threshold. If so, the evidence in the current case does not need to be "beyond a reasonable doubt" in order for this juror to conclude that the total evidence of guilt is beyond a reasonable doubt. Needless to say, many potential jurors had problems understanding the attorney's argument and were weeded out of the jury because it seemed that they would be influenced by past crimes—even though past crimes should have influenced them!

Jay heard that in another case, a prospective juror was asked "Do you think the defendant is more likely to be guilty just because he was arrested?" The juror's response was, "Of course! Unless you're arresting people at random." Spoken like a true Bayesian.

Patterns and Posteriors

The same logic applies to an assessment of patterns unearthed by data mining. Suppose that we are using a data mining algorithm to find a pattern that can be used to make predict stock prices, treat an illness, or serve some other worthwhile purpose. We use a reliable statistical test that will correctly identify a real pattern as truly useful and a coincidental pattern as truly useless ninety-five percent of the time.

We know that there are lots of useless patterns out there compared to the truly useful ones. So, let's say that one out of every 1,000 patterns that might be found by data mining *is* useful and the other 999 are useless. Our prior probability, before we find the pattern and do the statistical test, is one in a thousand. After we have found a pattern and determined that it is statistically significant, the posterior probability that it is useful works out to be less than one in 500. This is higher than one in 1,000, but it is hardly persuasive. We are still far more likely than not to have discovered a pattern that is genuinely useless.

This is a stark example of how probabilities and inverse probabilities can be quite different. The probability that a player in the English Premier League is male is 100 percent, but the probability that a randomly selected male plays in the English Premier League is close to 0. Here, even though there is only a five percent chance that a useless pattern will test statistically significant, more than ninety-eight percent of the patterns that test statistically significant are useless.

A one-in-a-thousand prior probability is surely too optimistic when data mining big data. Table E.2 shows the posterior probabilities for other values of the prior probability.

Table E.2 *Probability that a discovered pattern is useful.*

Prior Probability	Posterior Probability if Statistically Significant
0.001	0.018664
0.0001	0.001897
0.00001	0.000190
0.000001	0.000019

We don't know precisely how many useless patterns are out there waiting to be discovered, but we do know that with big data and powerful computers, it is a very large number, and that the number is getting larger every day. Chapter 5 recounts the story of the student who was sent on a wild goose chase to predict the exchange rate between the Turkish lira and U.S. dollar. With little effort, he looked at more than seventy-five million worthless patterns.

Our point is cautionary. Our distant ancestors survived and thrived because they recognized useful patterns in their environment. Today, the explosion in the number of things that are measured and recorded can provide us with useful information, but it has also magnified beyond belief the number of coincidental patterns and bogus statistical relationships waiting to deceive us.

There are now approximately forty trillion gigabytes of data, which is forty times more bytes than the number of stars in the known universe, and this data creation shows no signs of slowing down. Ninety percent of all the data that have ever been collected have been created in the past two years.

The number of possible patterns is virtually unlimited and the power of big computers to find them is astonishing. Since there are a relatively few useful patterns, most of what is found is rubbish—spurious correlations that may amuse us, but should not be taken seriously.

Don't be fooled by phantom patterns.

BIBLIOGRAPHY

"2010–2019 NBA All Star 3-Point Contests." Accessed September 29, 2019. https://www.justallstar.com/contests/3point/3p-2019-2018-2017-2016-2015-2014-2013-2012-2011-2010/#1548011618477-20160cea-ddf95d9f-c82f.

Abney, Wesley. "Live from Washington, It's lottery night 1969!", *Vietnam*, History Net. 2010. https://www.historynet.com/live-from-dc-its-lottery-night-1969.htm

Anderson, Chris. "The end of theory: The data deluge makes the scientific method obsolete", *Wired*, June 23, 2008. https://www.wired.com/2008/06/pb-theory/.

Anderson, Christopher, J., et al. "Response to comment on 'Estimating the reproducibility of psychological science'", *Science* 351 (2016): 1037.

Anderson, Jeff, Smith, Gary. "A great company can be a great investment." *Financial Analysts Journal* 62, no. 4 (2006): 86–93.

Angrist, Joshua, D. "Lifetime earnings and the Vietnam era draft lottery: Evidence from Social Security administrative records", *The American Economic Review* 80, no. 3 (1990): 313–36.

Asmelash, Leah, Muaddi, Nadeem. "Baby born on 7-Eleven Day at 7:11 p.m., weighing 7 pounds and 11 ounces." *CNN* online. July 13, 2019. https://edition.cnn.com/2019/07/13/us/711-baby-trnd/index.html.

Athalye, Anish, Engstrom, Logan, Ilyas, Andrew, Kwok, Kevin. "Fooling neural networks in the physical world with 3D adversarial objects." labsix. October 31, 2017. https://www.labsix.org/physical-objects-that-fool-neural-nets/

Athey, Susan. "The Impact of Machine Learning on Economics", in The Economics of Artificial Intelligence: An Agenda, edited by Ajay Agrawal, Joshua Gans, Avi Goldfarb, 507–52. Chicago, IL: University of Chicago Press, 2018.

Aubrey, Allison, "Can a pregnant woman's diet affect baby's sex?". *Morning Edition*, NPR. January 15, 2009. https://www.npr.org/templates/story/story.php?storyId=99346281&t=1580054221847

Baker, Monya, "1,500 scientists lift the lid on reproducibility." *Nature* 533, no. 7604 (2017): 452–4.

Bayer, Dave, and Diaconis, Persi. "Trailing the dovetail shuffle to its lair". *Annals of Applied Probability* 2, no. 2 (1992): 294–313.

Begoli, Edmonn, Horey, James. "Design principles for effective knowledge discovery from big data". Paper presented at Software Architecture (WICSA) and European Conference on Software Architecture (ECSA), 2012 Joint Working IEEE/IFIP Conference, Helsinki, August 2012.

Bem, Daryl, J. "Feeling the Future: Experimental Evidence for Anomalous Retroactive Influences on Cognition and Affect". *Journal of Personality and Social Psychology* 100, no. 3 (2011): 407–25.

Bhattacharjee, Yudhijit. "The mind of a con man". *The New York Times Magazine*, April 24, 2013.

Bode, Johann Elert. *Deutliche Anleitung zur Kenntniß des gestirnten Himmels [Clear Instruction for the Knowledge of the Starry Heavens]*. Hamburg: Dietrich Anton Harmsen, 1772.

Bohannon, J. "Replication effort provokes praise—and "bullying" charges". *Science* 344 (2014): 788–9.

Bolen, Johan, Mao, Huina, Zeng, Xiaojun. "Twitter mood predicts the stock market". *Journal of Computational Science* 2, no. 1 (2011): 1–8.

Borjas, G. J. "Ethnic capital and intergenerational mobility". *Quarterly Journal of Economics* 107, no. 1 (1992): 123–50.

Borjas, G. J. "The intergenerational mobility of immigrants", *Journal of Labor Economics* 11, no. 1 (1993): 113–35.

Brown, Nicholas, J. L., Coyne, James C. "Does Twitter language reliably predict heart disease? A commentary on Eichstaedt et al. (2015a)". *PeerJ* 6 (2018): e5656. https://doi.org/10.7717/peerj.5656.

Camerer, Colin, F., et al. "Evaluating replicability of laboratory experiments in economics". *Science* 351 (2016): 1433–6.

Camerer, Colin, F., et al. "Evaluating the replicability of social science experiments in Nature and Science between 2010 and 2015". *Nature Human Behaviour* 2, no. 9 (2018): 637–44.

Carliner, G. Wages, earnings, and hours of first, second, and third generation American males, *Economic Inquiry* 18, no. 1 (1980): 87–102.

Carrier, David, R., Kapoor, A. K., Tasuku Kimura, Nickels, Martin K., Scott, Eugenie C., So, Joseph K., Trinkaus, Erik. "The energetic paradox of human running and hominid evolution [and comments and reply]". *Current Anthropology* 25, no. 4 (1984): 483–95. https://doi.org/10.1086/203165

CBS News online. "Trump booed at World Series and chants of 'Lock him up!' break out". October 29, 2019. https://www.cbsnews.com/news/trump-booed-world-series-lock-him-up-chant-break-out-during-game-5-washington-nationals-stadium-pa/

Chambers, Sam. "The one thing Tesco's big data algorithm never predicted". *Bloomberg* online, October 20, 2015.

Chappell, Bill. "U.S. income inequality worsens, widening to a new gap." *NPR*. September 26, 2019. https://www.npr.org/2019/09/26/764654623/u-s-income-inequality-worsens-widening-to-a-new-gap.

Chatterjee, Rhitu. "New clues emerge in centuries-old Swedish shipwreck". *PRI*, February 23, 2012. https://www.pri.org/stories/2012-02-23/new-clues-emerge-centuries-old-swedish-shipwreck.

Chiswick, B. R. 1977. Sons of immigrants: are they at an earnings disadvantage?, *American Economic Review: Papers and Proceedings*, 67(1), 376–80.

Choi, Hyuntoung. 2012. Predicting the Present with Google Trends, *The Economic Record*, 88, special issue, 2–9.

Chordia, Tarun, Goyal, Amit, and Saretto, Alessio. "p-hacking: Evidence from two million trading strategies". Swiss Finance Institute Research Paper No. 17–37. Zurich: Swiss Financial Institute, 2017.

Christian, Brian. "The A/B Test: Inside the technology that's changing the rules of business". *Wired*, April 25, 2012.

Cios, Krzysztof, J., Pedrycz, Witold, Swiniarski, Roman W., Kurgan, Lukasz Andrzej. *Data Mining: A Knowledge Discovery Approach*, New York: Springer, 2007.

Climate Research Board. *Carbon Dioxide and Climate*. Washington, D.C.: National Academies Press, 1979. https://doi.org/10.17226/19856.

Coase, Ronald. "How should economists choose?", in *Ideas, Their Origins and Their Consequences: Lectures to Commemorate the Life and Work of G. Warren Nutter*, Washington, D.C: American Enterprise Institute for Public Policy Research, 1988.

Courtland, Rachel. "Gordon Moore: The man whose name means progress. The visionary engineer reflects on 50 years of Moore's Law". *IEEE Spectrum online*. March 30, 2015. https://spectrum.ieee.org/computing/hardware/gordon-moore-the-man-whose-name-means-progress.

Crane, Burton. *The Sophisticated Investor*. New York: Simon and Schuster, 1959, 56.

Docquier, Frédéric. "Income distribution, non-convexities and the fertility-income relationship". *Economica* 71, no. 282 (2004): 261–73.

Dunn, Peter, M. "James Lind (1716–94) of Edinburgh and the treatment of scurvy". *Archives of Disease in Childhood* 76 (1997): F64–5.

Egami, Naoki, Fong, Christian J., Grimmers, Justin, Roberts, Margaret E., Stewart, Brandon M. "How to make causal inferences using text". 2018. arXiv:1802.02163v12016.

Engber, Daniel. "Daryl Bem proved ESP is real: Which means science is broken". *Slate*, May 17, 2017.

Enserink, Martin. "Final report: Stapel affair points to bigger problems in social psychology". *Science*, November 28, 2012.

Fairley, Richard, E. Why the Vasa sank: 10 Lessons Learned". *IEEE Software online* 20, no. 2 (2003): 18–25.

Fayyad, Usama, Piatetsky-Shapiro, Gregory, Smyth, Padhraic. "From data mining to knowledge discovery in databases". *AI Magazine* 17, no. 3 (1996): 37–54.

Fink, Sheri. "This high-tech solution to disaster response may be too good to be true". *The New York Times*, August 9, 2019.

Fisher, Philip, A. *Common Stocks and Uncommon Profits*. Hoboken, NJ: John Wiley & Sons, 1958. Reprinted and updated 2003.

Garen, Brenton. "Artificial intelligence powered ETF debuts on NYSE". *ETF Trends*, October 18, 2017.

Gelsi, Steve, "Veterinary IPO barking in market". *MarketWatch* online, November 8, 2001. https://www.marketwatch.com/story/veterinary-firm-to-trade-under-woof-ticker

Gilbert, Daniel, T., King, Gary, Pettigrew, Stephen, Wilson, Timothy. D. "Comment on 'Estimating the reproducibility of psychological science' ". *Science*, 351 (2016): 1037.

Glaeser, Edward, Huang, Wei, Ma, Yueran, and Shleifer, Andrei. "A real estate boom with Chinese characteristics". *Journal of Economic Perspectives* 31, no. 1 (2017): 93–116.

GlobalPropertyGuide, "Home price trends". https://www.globalpropertyguide.com/home-price-trends/China. 2019.

Gray, Jim. "eScience—A Transformed Scientific Method". n.d. http://research.microsoft.com/en-us/um/people/gray/talks/NRC-CSTB_eScience.ppt

Gray, Paul, E. "The man who loved children: Dr. Benjamin Spock (1903–1998)". *Time* 151, no. 12. March 30, 1998. http://content.time.com/time/magazine/article/0,9171,988061,00.html.

Harish, Ajay. "Why the Swedish Vasa ship sank", *SimScale* (blog), March 21, 2019. https://www.simscale.com/blog/2017/12/vasa-ship-sank/.

Heinonen, O. P., Shapiro, S., Tuominen, L., Turunen, M. I. "Reserpine use in relation to breast cancer". *Lancet* 2 (1974): 675–7.

Hennessey, Raymond. "VCA Antech has a dismal debut as it makes second public run". *Dow Jones Newswires* online. November 23, 2001. https://www.wsj.com/articles/SB1006372433208957200.

Hernandez, Ben. "You'll 'be at a disadvantage' without AI, says Tom Lydon on Yahoo! Finance LIVE". *ETF Trends* online. March 25, 2019.

Herndon, Thomas, Ash, Michael, Pollin, Robert. "Does high public debt consistently stifle economic growth? A critique of Reinhart and Rogoff". *Cambridge Journal of Economics* 38, no. 2 (2014): 257–79.

Kahneman, Daniel, Tversky, Amos. "Prospect theory: An analysis of decision under risk". *Econometrica* 47, no. 2 (1979): 263–92.

Inskeep, Steve. "Missouri baby born on July 11, which is known as 7–11 Day". Morning Edition: Strange News, *NPR*. July 15, 2019.

Ioannidis, John, A. "Contradicted and initially stronger effects in highly cited clinical research". *Journal of the American Medical Association* 294, no. 2 (2005): 218–28.

Jaki, Stanley, L. The Early History of the Titius-Bode Law, *American Journal of Physics* 40, no. 7 (1972): 1014. https://doi-org.ccl.idm.oclc.org/10.1119/1.1986734

John, Leslie, K., Loewenstein, George, Prelec, Drazen. "Measuring the prevalence of questionable research practices with incentives for truth-telling". *Psychological Science* 23, no. 5 (2012): 524–32.

Jones Larry, E., Schoonbroodt Alice, Tertilt, Michèle. *Fertility theories: Can they explain the negative fertility-income relationship?* NBER Working Paper 14266. Cambridge, MA: National Bureau of Economic Research, 2008.

Jordan, Brigitte Jordan. "Pattern Recognition in Human Evolution and Why It Matters for Ethnography, Anthropology, and Society", in *Advancing Ethnography in Corporate Environments: Challenges and Emerging Opportunities*, edited by Brigitte Jordan, 193–214. Abingdon: Routledge, 2016.

Karelitz, Samuel, Fisichelli, Vincent R., Costa, Joan, Kavelitz, Ruth, Rosenfeld, Laura. "Relation of crying in early infancy to speech and intellectual development at age three years". *Child Development* 35 (1964): 769–77.

Kecman Vojislav. "Foreword", in *Data Mining: A Knowledge Discovery Approach*, edited by Krzysztof J. Cios, Witold Pedrycz, Roman W. Swiniarski, Lukasz Andrzej Kurgan, xi. New York, Springer, 2007.

"Kenyans chase down and catch goat-killing cheetahs". *BBC* online. November 15, 2013. https://www.bbc.com/news/world-africa-24953910.

Knapp, Gunnar. *Why You'll Never Get Elected Congressman from the Bronx Unless Your Last Name Starts with B.* unpublished thesis, Yale University, December 7, 1973.

LeClaire, Jennifer. "Moore's Law Turns 50, Creator Says It Won't Go on Forever". *Newsfactor*, May 12, 2015.

Lee, Jane, J. "How a rooster knows to crow at dawn". *National Geographic News*, March 19, 2013. https://www.nationalgeographic.com/news/2013/3/130318-rooster-crow-circadian-clock-science/.

Levelt Committee, Noort Committee, Drenth Committee. *Flawed science: The fraudulent research practices of social psychologist Diederik Stapel*. November 28, 2012. https://poolux.psychopool.tu-dresden.de/mdcfiles/gwp/Reale%20F%C3%A4lle/Stapel%20-%20Final%20Report.pdf.

Levy, Steven. "Exclusive: How Google's algorithm rules the web". *Wired*, February 22, 2010.

Lévi-Strauss, Claude. Tristes Tropiques. Translated by John Weightman and Doreen Weightman. New York: Penguin Books, 1992.

Lewis, Tanya. "Gorgeous Decay: The Second Death of the Swedish Warship *Vasa*". *Wired*, September 3, 2012. https://www.wired.com/2012/09/swedish-warship-vasa/.

Liebenberg, Louis. "The relevance of persistence hunting to human evolution, *Journal of Human Evolution* 55, no. 6 (2008): 1156–9.

Liu, Yukun, Tsyvinski, Aleh. "Risks and returns of cryptocurrency". NBER Working Paper No. w24877. August 6, 2018. https://ssrn.com/abstract=3226952.

Lo, Bruce, M., Visintainer, Catherine M., Best, Heidi A., Beydoun, Hind A. "Answering the myth: Use of emergency services on Friday the 13th". *The American Journal of Emergency Medicine* 30, no. 6 (2012): 886–9.

Macaskill, Sandy. "Top 10: Football superstitions to rival Arsenal's Kolo Toure". *The Telegraph* February 25, 2009.

Man, Joyce Yanyun, Zheng, Siqi, Ren, Rongrong. "Housing Policy and Housing Markets: Trends, Patterns, and Affordability", in *China's Housing Reform and Outcomes*, edited by Joyce Yanyun Man, 3–18. Cambridge, MA: Lincoln Institute of Land Policy, 2011.

Mann, Brad. "How many times should you shuffle a deck of cards?". *UMAP* 15, no. 4 (1994): 303–32.

Markowitz, Harry. Portfolio Selection: Efficient Diversification of Investments. New York: John Wiley & Sons, 1959.

Marr, Bernard. "Big Data at Tesco: Real Time Analytics at The UK Grocery Retail Giant." *Forbes*, November 17, 2016. https://www.forbes.com/sites/bernardmarr/2016/11/17/big-data-at-tesco-real-time-analytics-at-the-uk-grocery-retail-giant/#7196f50061cf.

Mathews, Fiona, Johnson, Paul J., Neil, Andrew. "You are what your mother eats: Evidence for maternal preconception diet influencing foetal sex in humans". *The Royal Society: Proceedings: Biological Sciences* 275, no. 1643 (2008): 1661–8.

Mattson, Mark, P. "Superior pattern processing is the essence of the evolved human brain". *Frontiers in Neuroscience* 8 (2014): 265. doi:10.3389/fnins.2014.00265

Mayer, Jürgen, Khairy, Khaled, Howard, Jonathon. "Drawing an elephant with four complex parameters". *American Journal of Physics* 78 (2010): 648. DOI:10.1119/1.3254017.

McMahon, Dinny. *China's Great Wall of Debt: Shadow Banks, Ghost Cities, Massive Loans, and the End of the Chinese Miracle*. Boston, MA: Houghton Mifflin Harcourt, 2018.

McRae, Mike. "Science's 'replication crisis' has reached even the most respectable journals, report shows." ScienceAlert. August 27, 2018. https://www.sciencealert.com/replication-results-reproducibility-crisis-science-nature-journals.

Meier, Paul. "The Biggest Public Health Experiment Ever: The 1954 Field Trials of the Salk Poliomyelitis Vaccine", in *Statistics: A Guide to the Unknown*, edited by Judith Tanur, Frederick Mosteller, William H. Kruskal, Richard F. Link, Richard S. Pieters, Gerald R. Rising, 2–13. San Francisco: Holden-Day, 1972.

Milbank, Dana. "President Trump is entering his terrible twos". *The Washington Post*, January 7, 2019.

Mitchell, Robert, L. "12 predictive analytics screw-ups". *Computerworld* online, July 24, 2013. https://www.computerworld.com/article/2484224/12-predictive-analytics-screw-ups.html

Moore, Geoffrey. "Cramming more components onto integrated circuits". *Electronics* 38, no. 8, April 19, 1965: 114.

Mullen, Jethro, Stevens, Andrew. Billionaire: Chinese real estate is "biggest bubble in history". CNN Business, *CNN* online. September 29, 2016. https://money.cnn.com/2016/09/28/investing/china-wang-jianlin-real-estate-bubble/

Myers, D. "Upward mobility in space and time: lessons from immigration", in *America's Demographic Tapestry*, edited by James W. Hughes, Joseph J. Seneca, 135–57. New Brunswick, NJ: Rutgers University Press, 1999,

Neill, Tess. Tesco's foray, and failure, in the U.S., *AdAge* October 4, 2013s.

Nelson, Leland, R., Furst, Merrick L. "An objective study of the effects of expectation on competitive performance". *The Journal of Psychology* 81 (1972): 69–72.

Nicholls, Neville. "The Charney Report: 40 years ago, scientists accurately predicted climate change." Phys.org online. July 23, 2019. https://phys.org/news/2019-07-charney-years-scientists-accurately-climate.html.

Nisbet, Robert, Elder, John, Miner, Gary. *Handbook of Statistical Analysis & Data Mining Applications*. Burlington, MA: Academic Press, 2009.

Nunlist, Tom. "China's housing market: The iron bubble". *Cheung Kong Graduate School of Business Knowledge* online. July 31, 2017.

Open Science Collaboration. "Estimating the reproducibility of psychological science". *Science* 349, no. 6251 (2015). DOI: 10.1126/science.aac4716.

Palca, Joe. "Betting on artificial intelligence to guide earthquake response". All Things Considered, *NPR*, April 20, 2018.

Pashler, Harold, Wagenmakers, Eric. "Editors' Introduction to the special section on 'Replicability in psychological science: A crisis of confidence?'". *Perspectives on Psychological Science* 7, no. 6 (2012): 528–30.

Peirce, Charles Sanders, Ketner, Kenneth Laine. *Reasoning and the logic of things: The Cambridge conferences lectures of 1898*. Cambridge, MA: Harvard University Press, 1992, 194–6.

Piatetsky-Shapiro, G. "Knowledge discovery in real databases: A report on the IJCAI-89 workshop". *AI Magazine* 11, no. 5 (1991): 68–70.

Powell, Adam, "Humans hot, sweaty, natural-born runners". *The Harvard Gazette* April 19, 2007.

Preis, Tobias, Moat, Helen Susannah, Stanley, H. Eugene. "Quantifying trading behavior in financial markets using Google trends". *Scientific Reports* 3 (2013): 1684.

Rashes, Michael. "Massively confused investors making conspicuously ignorant choices". *The Journal of Finance* 56 (2001): 1911–27.

Redman, Thomas, C. "Bad data costs the U.S. $3 trillion per year". *Harvard Business Review*, September 22, 2016.

Roman, John, Chalfin, Aaron. *Is there an iCrime Wave?* Washington, D.C.: Urban Institute, 2007.

Sadrô, J., Janudi, Izzat, Sinha, Pawan. "The role of eyebrows in face recognition". *Perception* 32, no. 3 (2003): 285–93.

Sagiroglu, S, Sinanc, D. "Big data: A review". Paper presented at The 2013 International Conference on Collaboration Technologies and Systems (CTS 2013), San Diego, CA, May 2013.

Scanlon, Thomas, J., Luben, Robert, N., Scanlon, F. L., Singleton, Nicola. "Is Friday the 13th bad for your health?". *British Medical Journal* 307, no. 6919 (1993): 1584–6.

Schrage, Michael. 2014. Tesco's Downfall Is a Warning to Data-Driven Retailers, *Harvard Business Review*, October.

Schuld, Jochen, Slotta, Jan E., Schuld, Simone, Kollmar, Otto, Schilling, Martin K., Richter, Sven. "Popular belief meets surgical reality: Impact of lunar phases, Friday the 13th and zodiac signs on emergency operations and intraoperative blood loss". *World Journal of Surgery* 35, no. 9 (2011): 1945–9.

Schwab, Katherine. "Disaster relief is dangerously broken. Can AI fix it?". *Fast Company* online, November 15, 2018.

Schwab, Katherine. "This celebrated startup vowed to save lives with AI. Now, it's a cautionary tale". *Fast Company* online, August 13, 2019.

Shanks, David, R., Newell, Ben R., Lee, Eun Hee, Balikrishnan, Divya, Ekelund, Lisa, Cenac, Zarus, Kavvadia, Fragkiski, Moore, Christopher. "Priming intelligent behavior: An elusive phenomenon". *PLOS One* 2013. https://doi.org/10.1371/journal.pone.0056515.

Shapiro, Samuel. "Looking to the 21st century: Have we learned from our mistakes, or are we doomed to compound them?". *Pharmacoepidemiology and Drug Safety* 13, no. 4 (2004): 257–65.

Shapiro, S., Slone, D. "Case-control study: Consensus and controversy". *Comment, Journal of Chronic Disease* 32 (1979): 105–7.

Sharpe, William F. "Mutual fund performance". *Journal of Business* January (1966): 119–38.

Simmons, Joseph P., Nelson, Leif D., Simonsohn, Uri. "False-positive psychology: undisclosed flexibility in data collection and analysis allows presenting anything as significant". *Psychological Science* 22 (2011): 1359–66.

Simonite, Tom, "Moore's law is dead. Now what?". *MIT Technology Review*, May 13, 2016. https://www.technologyreview.com/s/601441/moores-law-is-dead-now-what/.

Simonsohn, Uri. "Just post it: The lesson from two cases of fabricated data detected by statistics alone". SSRN 2012. https://doi.org/10.2139/ssrn.2114571.

Smith, Gary. "Great company, great investment revisited". *Journal of Wealth Management* 19, no. 1 (2016): 34–9.

Smith, Gary. *The AI Delusion*. Oxford: Oxford University Press, 2018.

Smith, Gary, Levere, Michael, Kurtzman, Robert. "Poker player behavior after big wins and big losses". *Management Science* 55, no. 9 (2009): 1547–55.

Smith, Gary. *The Paradox of Big Data*. Unpublished, 2019.

Smith, Gary, Cordes, Jay. *The 9 Pitfalls of Data Science*. Oxford: Oxford University Press, 2019.

Smith, Margaret Hwang, and Smith, Gary. "Bubble, bubble, where's the housing bubble?". *Brookings Papers on Economic Activity* 2006, no. 1 (2006): 1–50.

Smith, Margaret Hwang, Smith, Gary. "Like mother, like daughter?: An economic comparison of immigrant mothers and their daughters". *International Migration* 51 (2013): 181–90.

Stein, Gregory, M. "What will china do when land use rights begin to expire?". *Vanderbilt Journal of Ttransnational Law* 50 (2017): 625–72.

Su, Francis, E., et al. "Seven shuffles". *Math Fun Facts*. https://math.hmc.edu/funfacts/seven-shuffles/. n.d.

Thelwell, Richard. "The dog that didn't bark: How did customer analytics go so badly wrong at Tesco?:. https://www.matillion.com/insights/how-did-customer-analytics-go-so-badly-wrong-at-tesco/.

Tullock, Gordon. "A comment on Daniel Klein's 'A plea to economists who favor liberty'". *Eastern Economic Journal* 27, no. 2 (2001): 203–7.

Turner, Annie. "The cautionary tale of Tesco: Big data and the 'analytic albatross'". *Inform online*. January 2015. https://inform.tmforum.org/features-and-analysis/2015/01/cautionary-tale-tesco-big-data-analytic-albatross/.

Tversky, Amos, Kahneman, Daniel. "On the psychology of prediction". *Psychological Review* 80 (1973): 237–51.

Tversky, Amos, Kahneman, Daniel. "Judgement under uncertainty: Heuristics and biases". *Science* 185 (1974): 1124–31.

Ullman, S. B. "A mathematical formulation of Bode's law". *Popular Astronomy*, 57 (1949): 197.

Urban Institute. *The iPod: Lightning rod for criminals?*, Washington, D.C.: Urban Institute, September 27, 2007.

Vogelstein, Fred. How Yahoo! blew it, *Wired*, February 1, 2007.

Wei, Shang-Jin, Zhang, Xiaobo, Liu, Yin. "Home ownership as status competition: Some theory and evidence". *Journal of Development Economics* 127 (2017): 169–86.

Weinberg, Robert, S., Gould, Daniel, Yukelson, David, and Jackson, Allen. "The effect of preexisting and manipulated self-efficacy on a competitive muscular endurance task". *Journal of Sport and Exercise Psychology* 3, no. 4 (1981): 345–54.

Wicherts, Jelte, M., Bakker, Marjan, Molenaar, Dylan. "Willingness to share research data is related to the strength of the evidence and the quality of reporting of statistical results". *PLoS ONE* 6, no. 11 (2011). https://doi.org/10.1371/journal.pone.0026828.

Worland, Justin, "How we can stop earthquakes from killing people before they even hit". *Time* September 28, 2017.

Wu, Jing, Deng, Yongheng, Liu, Hongyu. "House price index construction in the nascent housing market: The case of China". *The Journal of Real Estate Finance and Economics* 48, no. 3 (2014): 522–45.

Young, Shalise Manza. "NFL, NFLPA say there's 'no evidence' Panthers' Eric Reid was targeted for drug testing". *Yahoo Sports*, January 9, 2019. https://sports.yahoo.com/nfl-nflpa-say-theres-no-evidence-panthers-eric-reid-targeted-drug-testing-150426603.html

Young, Stanley, S., Bang, Heejung, Oktay, Kutluk. "Cereal-induced gender selection? Most likely a multiple testing false positive". *The Royal Society: Proceedings: Biological Sciences* 276, no. 1660 (2009): 1211–12.

Zeisel, Hans. "Dr. Spock and the case of the vanishing women jurors". *University of Chicago Law Review* 37, no. 1 (1969): 1–18.

INDEX